十年十筑

赵春水 著

江苏凤凰科学技术出版社·南京

图书在版编目（CIP）数据

十年十筑 / 赵春水著. -- 南京：江苏凤凰科学技术出版社, 2022.10
　ISBN 978-7-5713-2775-0

Ⅰ. ①十… Ⅱ. ①赵… Ⅲ. ①城市建筑 – 建筑设计 – 研究 – 天津 Ⅳ. ① TU984.221

中国版本图书馆 CIP 数据核字 (2022) 第 028640 号

十年十筑

著　　者	赵春水
译　　者	刘晓烨　陈　旭　张　萌
项目策划	凤凰空间 / 陈　景
责任编辑	赵　研　刘屹立
特约编辑	狄　阙　田　园
出版发行	江苏凤凰科学技术出版社
出版社地址	南京市湖南路 1 号 A 楼，邮编：210009
出版社网址	http://www.pspress.cn
总 经 销	天津凤凰空间文化传媒有限公司
总经销网址	http://www.ifengspace.cn
印　　刷	天津图文方嘉印刷有限公司
开　　本	710 mm×1000 mm　1 / 16
印　　张	15
字　　数	280 000
版　　次	2022 年 10 月第 1 版
印　　次	2022 年 10 月第 1 次印刷
标准书号	ISBN 978-7-5713-2775-0
定　　价	128.00 元

图书如有印装质量问题，可随时向销售部调换（电话：022-87893668）。

前言

2009—2018年，约3650个日出日落，我们见证了十个建筑从无到有、从概念到现实的历程。期间审慎地研究每一个项目，不断自我否定挑战极限，因为我们不仅要实现"建筑改变社会"的理想，更希望通过我们设计的建筑为社会的进步做些许贡献，同时能传递给人们应有的尊重和关爱。

十年对一个城市的历史来讲不算长，但这十年中，天津的城市结构和形态却发生了史无前例的巨变。2004年海河两岸的改造拉开了天津21世纪城市建设的序幕，打下了跻身国际都市的基础。2008年北京奥运会后，天津的城市建设日新月异，进入快速发展时期。作为这些变化的参与者与见证者，我们挑选了由自己设计的十个建筑项目同大家分享。本书记录了本土设计师不断成长的心路历程，有应对挑战的成功与失败，更有我们对城市、对建筑的理想与坚守。我们坚持从城市的角度去研究建筑，从社会的角度去思考建筑，从人文的角度去体验建筑，这些成为我们十年来坚持的创造理念。

为方便读者阅读，本书将十个建筑作品按两类进行归纳叙述，前六个是继承传统风格的作品，后四个是创造新空间、描绘现代城市生活的作品。

天津建卫600余年，九河下梢、华洋杂糅的历史决定了其城市基因中具有独特的气质。传承并发展是对历史街区最有效的保护，泰安道二号院、四号院的建设，为天津乃至全国的历史文化街区复兴提供了有参考价值的经验。原租界上保留的大量遗存建筑成为天津城市建筑的底色，中西合璧、古今兼容的西方建筑群也给人们留下了深刻的印象。天津职业大学是我们自主选择延续天津地域性和传承教育建筑类型的尝试。对于天津市民族文化宫项目，尽管我们做了许多有独立见解的方案来表现继承与发展的关系，但出于稳妥起见最终保守地为天津留下一个合理且让人熟悉的结果。功能分区明确和机动车优先的城市规划理念让天津同其他许多城市一样失去了城市空间的连续性，使自由步行成为奢望，但黑牛城道八大里社区的出现改变了这一现状，该区域已成为全国第一个实现"窄路密网、开放街区、混合社区"的新型居住区典范，其适老住宅也使普通商品房自带高端人文气质。天津市第四中学作为该区域配套的完全中学，以创造交流空间和共享教育资源为设计目标，实现了传统材料在极限场地条件下新教育空间的建构。

天津在历史上是中国北方受西方影响最深的城市，在近代也是建筑文化最时尚的北方万国建筑博览会，城市底蕴中蓄藏着对新风格的接纳和包容。海河改造开创了现代天津在设计领域大规模"东学西鉴"的先河，更是中国设计界改革和发展的一个缩影，随后天津文化中心建设成为中外建筑师竞赛的舞台。天津图书馆在为城市提供公共空间的同时，将建造技术与空间艺术完美结合，创造出前所未有的室内空间，成为图书馆发展模式的里程碑。滨海图书馆在强调延续城市公共空间属性的同时，积极应对人们阅读方式的转变，创造将时尚造型与丰富寓意融为一体的"滨海之眼"，成为滨海新区的文化地标。滨海科技馆继承场地原有记忆，摆脱原功能消失的宿命，结合新空间和行为需求营造立体的"拉维莱特公园"，发扬公共性、场地性价值，赋予该区域新的历史记忆。随着这三个作品的完成并投入使用，它们的影响力唤回了天津城市建设曾经展示的丰富性与先进性，并将重新定位现代文化建筑的城市公共属性和社会历史属性。

　　济宁市群众艺术馆是我们竞标后独立完成的作品，也是《十年十筑》中唯一的外埠项目。我们为"孔孟之乡、运河之都"的济宁量身定制了文化中心3.0版本，进一步强化文商结合的发展模式。群众艺术馆源于《论语》"游于艺"的治学理想，设计中我们用先进的建造技术为群众创造了开放、流动的空间，尝试将"游于艺"的抽象概念用可感知的现实空间体验加以诠释。建成后，由于其提供了充分的交流场所和丰富的空间感受，是济宁文化中心使用率最高的场馆，为文化中心注入了活力，成为群众艺术馆建筑的新范式。

　　回顾过去的十年，中国的社区发生了许多变化，尤其是互联网技术带来的冲击使人们的生活发生了根本性改变。这些变化不应疏远我们同历史的距离，而是应该让我们深刻地认知、继承在现代互联网思维语境下的精神意义和社会价值。对城市来讲，重视世人的日常生活和身体感知，比单纯地追求形式的标志性和视觉愉悦更具有现实意义；对建筑来讲，继承传统文化必须同大众共享才能体现文化与道德的社会价值。在这个已经到来的扁平化世界中，由时代精神统领的趋势逐渐在减弱，更多的建筑理想主义者有了表达自己理解世界的机会，得以在更广阔的舞台上设计出向世人传递理解和关爱的空间，从而来改变世界。本书不仅是对过去十年工作的总结，更包含对未来建筑设计如何"继承和创新"的展望。外部世界存在着很多不确定性，我们只有坚守内心对建筑的那份确定性才是应对这些不确定性的唯一方法。

赵春水

2020年6月于黄埔南路

目录

泰安道二号院办公楼   6
Tai'an Road No. 2 Courtyard Office Building

丽思·卡尔顿酒店（泰安道四号院）   26
Ritz-Carlton Hotel (No. 4 Courtyard)

天津职业大学   50
Tianjin Vocational Institute

天津市民族文化宫   64
Tianjin Culture Palace of Nationalities

黑牛城道五福里   80
Wufu Community, Heiniucheng Road

天津市第四中学迁址扩建工程   102
Tianjin No.4 Middle School Relocation and Expansion Project

天津图书馆   128
TianJin Library

滨海图书馆   154
Binhai Library

滨海科技馆   178
Binhai Science and Technology Museum

济宁市群众艺术馆   208
Jining Mass Art Museum

设计团队名单   236
Team Members

历史街区的更新除了功能与业态的重新植入以外,在形态与风格方面一直存在两种不同的思考:一是以专业人士秉承的反映时代风格为目标,试图用现代的形式语言以区别传统风格来标识其时代性,但想象的形式往往由于现代建材及技术的粗陋失去其追求的先进性而成为传统建筑的陪衬;二是建立在大众对传统风格的认知之上,对传统形式进行简化,采用传统材料及工艺,结合具体功能创造性地协调各种既有条件建构可识别的公共性空间,使新旧协调不是基于简单对比,而是根植于对尺度和工艺的把握与传承。

泰安道二号院办公楼现为游艇俱乐部使用,就是采用"大众"的方法,通过技艺、材料以及手工建造的传承来统一新旧建筑重塑都市空间。地块内有天津第一饭店和原十八集团军驻津办事处等保留建筑,为了从体量、风格上协调各方形成一个整体,新建建筑限高24 m。立面上水平三段式、竖向五段式,分段方式与北侧丽思·卡尔顿酒店相呼应,首层设单层柱廊,协调道路尺度。

泰安道二号院办公楼
Tai'an Road No. 2 Courtyard Office Building

用地面积 | 11 856 m²
建筑规模 | 36 589 m²
建筑团队 | 天津市城市规划设计研究总院
泛光照明设计 | 天津市城市规划设计研究总院
获奖情况 | 2015年度"海河杯"天津市优秀勘察设计
建筑工程公建二等奖

实景 | scene

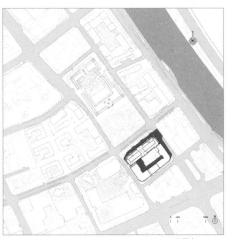

区位图 | location

新旧建筑共同围合院落，提供给城市珍贵的公共空间。新建部分不仅要负担自身的使用需求，还要解决既有老建筑因时代变迁在能源供给、消防安全、交通组织方面留下的隐患。兼顾新与旧将能源系统、消防系统和交通系统统一整合到地下空间，使原来被设备用房占据的内院成为面向都市的开放场所。精心呵护的建筑背面和内院空间使建筑外表和内在都得到有效提升，实现了真实意义上的更新，让新旧建筑共享技术进步带来的成果。

泰安道二号院办公楼的重塑既解决了老建筑配套设施落后、安全隐患严重的现实问题，又改造了建筑系统，形成体面的开放空间，兼顾尺度、工艺传承，采用现代技术使老建筑实现内在真实提升，主动营造的内院使传统建筑场所的复兴和新建筑的有机生长成为互为依托、彼此存在的前提。第三空间（院落空间）的创设为新旧建筑提供一个相互对话的公共广场，同时将空间要素整合，成为城市更新中应对新旧功能迭代的有效策略。

In addition to the re-implantation of functions and formats in the historic areas, there are always two different kinds of thinking in terms of form and style: First, the "professional" personnel adheres to the goal of reflecting the style of the times, trying to distinguish them with modern formal languages. The traditional style identifies its era, but the form of imagination often becomes the foil of traditional architecture due to the loss of the advanced nature of modern building materials and technology. The second is based on the recognition of traditional style by the "mass". The traditional form is simplified by using traditional materials and techniques, and creatively coordinating various existing conditions to construct an identifiable public space in combination with specific functions, so that the old and new coordination is not based on simple comparison, but is rooted in the grasp and inheritance of scales and processes.

The yacht club is a "mass" approach that unifies new and old buildings to reshape urban spaces through craftsmanship, materials and hand-built heritage. In the plot, there are reserved buildings such as the Tianjin First Hotel and the 18th Road Army Office, which form a whole body for the coordination of the volume and style. The new building has a height limit of 24m. The horizontal three-section vertical five-segment on the façade echoes the Ritz-Carlton hotel on the north side. The first floor has a single-level colonnade to coordinate the road scale.

New and old buildings are enclosed in courtyards and provide the city with precious public spaces. The new part not only has to bear its own needs, but also solves the hidden dangers of the old buildings due to the changes of the times in terms of energy supply, fire safety and traffic organization. Taking into account the new and old integration of the energy system, fire protection system and transportation system into the underground space, the inner courtyard originally occupied by the equipment room becomes an open place facing the city. The well-protected back of the building and the inner courtyard space enable the appearance and interior to be effectively upgraded to achieve a real sense of renewal, so that the new and old buildings share the fruits of technological progress.

The reshaping of the yacht club not only solves the practical problems of the backward security risks of the old building supporting facilities, but also transforms the building system to make the back of the city a decent open space, taking into account the scale and process inheritance, and adopting modern technology to realize the inner real improvement of the old building. The active inner courtyard has made the rejuvenation of traditional buildings and the organic growth of new buildings a prerequisite for mutual existence. The creation of the third space is an attempt to update the city, and it is also a strategic concept for the iteration of new and old functions.

实景 | scene

拆除前杂乱无序 | before demolition: disorganized space

保留建筑 | reserved buildings

拆除后绿地空旷 |
after demolition: green open space

建成后重塑都市空间 | after completion: reconstruction of urban space

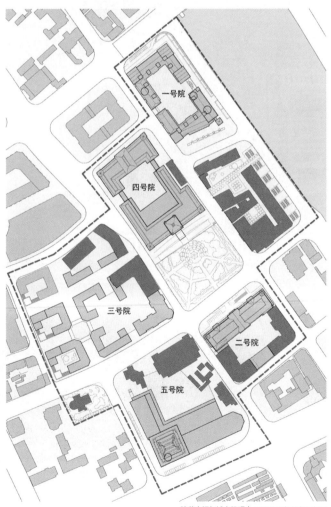

保留与新建相结合 |
reserved and new buildings

院落空间与城市肌理 | courtyard and texture

实景 | scene

沿泰安道立面图 | facade along Tai'an Road

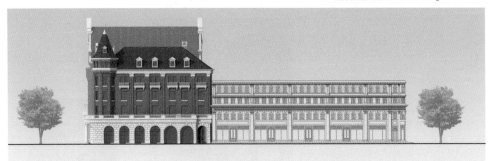

沿大沽北路立面图 | facade along North Dagu Road

沿解放北路立面图 | facade along North Jiefang Road

剖面图 | section

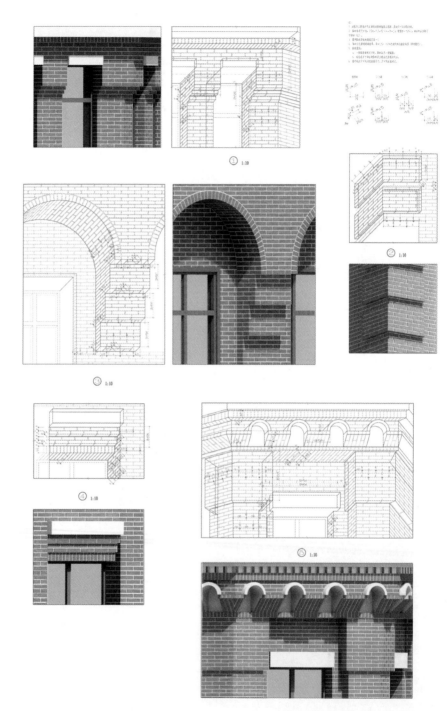

局部节点详图 | partial detail of structure

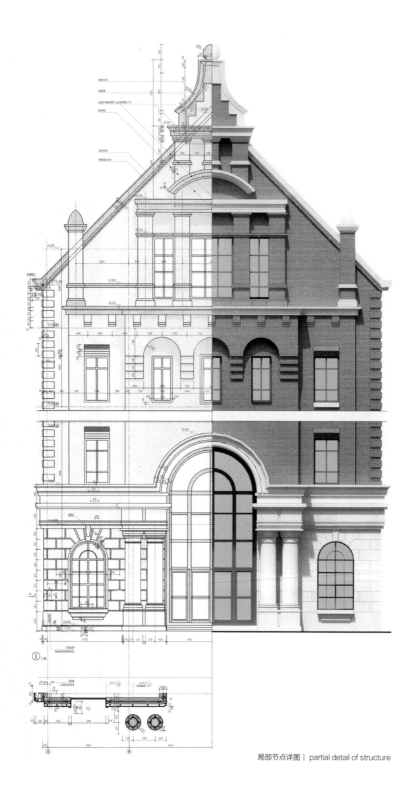

局部节点详图 | partial detail of structure

休息厅 | resting room

门厅 | hall

展示区局部 | partial show space

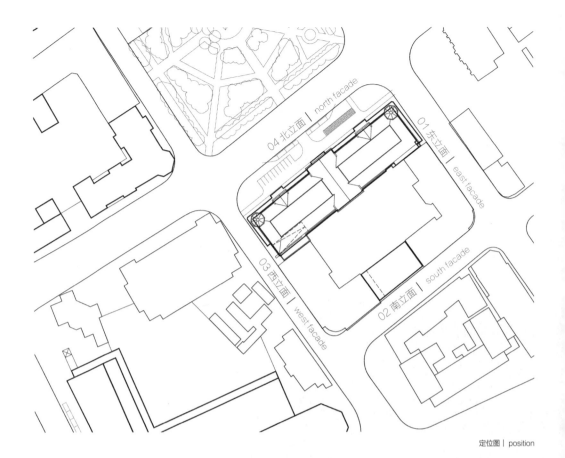

定位图 | position

外部东立面 | east facade

外部南立面 | south facade

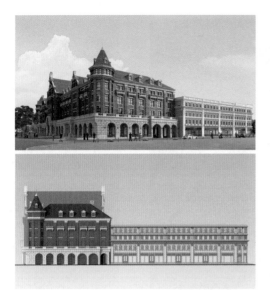

外部西立面 | west facade

外部北立面 | north facade

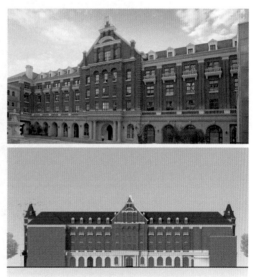

内部西立面 | west facade

内部北立面 | north facade

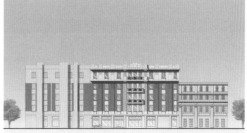

内部南立面 | south facade

主入口 | main entrance

主入口：门洞 | main entrance : doorway

主入口：通廊 | main entrance : corridor

设备用房：施工前 | equipment room: pre-construction

设备用房：竣工后 | equipment room: after completion

内院：施工前 | inner court : pre-construction

设备用房：整体迁到地下一层 | equipment room: basement levelone

内院：竣工后 | inner court : after completion

鸟瞰图 | aerial view

天津是一座有故事的城市,建卫600多年来给我们留下无数宝贵的精神和物质财富,以海河为主线,解放北路、五大道、意式风情区都向世人诉说着津沽大地曾经的辉煌与荣耀。但由于历史变迁保护区日渐衰落,对历史街区的保护和追求利益最大化的推倒重建之间的矛盾,困扰着试图再现津沽风华的人们,天津市政府的保护式搬迁为我们在现实与理想之间找到一条城市有机更新的道路提供了难得的机会。

泰安道地区位于有"北方华尔街"之称的解放北路与五大道之间,保留了16处极具价值的文物建筑,如戈登堂、安里甘教堂、利顺德饭店、第一饭店、开滦矿务局、美国兵营等,它们的存在彰显了该地区的历史人文价值。围绕有着百年历史的维多利亚花园,我们以恢复街区尺度为目标,以重塑院落为手段,以协调建筑风格为方法,开展了五个街区的规划和建筑设计工作。

历史街区的最大建筑高度为45 m,首层和二层总高12 m,顶层及屋顶高10 m,三段式立面设计兼顾传统与均衡。西侧、北侧、东侧保证建筑的贴线率,采用骑楼形式强化与街道的联系,在东西侧呼应城市道路留

丽思·卡尔顿酒店（泰安道四号院）
Ritz-Carlton Hotel (No. 4 Courtyard)

用地面积	17 483 m²
建筑规模	98 215 m²
建筑团队	天津市城市规划设计研究总院、丽思·卡尔顿酒店亚太地区设计部
内装设计	法国 PYR 设计公司、天津市建筑设计院、天津市城市规划设计研究总院
泛光照明设计	北京富润成照明系统工程有限公司、天津市城市规划设计研究总院
智能化设计	天津市天房科技发展股份有限公司
其他团队	天津致通机电设计事务所、德国戴水道（Atelier Dreiseitl）设计公司、美国 RND 设计公司、日本 Strickland 设计事务所
获奖情况	2015 年度"海河杯"天津市优秀勘察设计建筑工程公建一等奖 2014 年度天津市建设工程"金奖海河杯"奖

实景 | scene

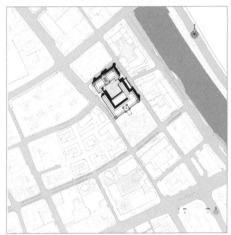

区位图 | location

出出入入口，使内部院落与街道连通并成为城市公共空间的重要一员。丽思·卡尔顿酒店（Ritz-Carlton）集团对品质和历史街区的氛围的独特理解，让我们确立"为客户提供回家的感受"为设计目标，在数万字"作业标准书"的指导下，从设计、运营和客户体验角度对门厅、宴会厅、客房、健身房等反复比较方案，以宜人尺度和舒适流线为客户提供优质体验。为保持同文物建筑的外沿质感协调，采用真砖砌筑，结合传统工艺，我们专门绘制石材、砌块、门窗和屋顶的三维大样详图来指导现场施工，现在看来这是保证完成度的一项极为重要的工作。

从 2015 年开业至今，丽思·卡尔顿酒店已成为万豪集团亚洲历史街区的旗帜性存在，给他们带来巨大利益和声誉。它的成功也鼓励着我们今后在面对历史街区更新改造中，注重尺度、形式和功能的同时更关注人的体验和氛围营造，坚定了我们在新都市文化背景下，用街区重塑理念和结合传统工艺的方法延续历史传承，为继续讲述天津自己的故事提供最佳场所和现实背景。

Tianjin is a city with stories. From its founding, the city has left us with countless precious spiritual and material wealth for 600 years. The Haihe River is the main line. North Jiefang Road, the Five Avenues(Wudadao) and the Italian style District all tell the world about the glory of the land of Tianjin. However, due to the gradual decline of historically changed protected areas, the contradiction between the protection of historical blocks and the rebuilding of the pursuit of maximum profits has plagued people trying to reproduce the glory of Tianjin. Finding a road to organic renewal in a city offers relocation which is between dream and reality.

The Tai'an Road area is located between North Jiefang Road and the Five Avenues, which is known as the "North Wall Street". It retains 16 valuable cultural relics such as the Gordon Hall, All Saint Church, the Lishunde Hotel, the First Hotel, Kailuan Mineral Board and the American Barracks, etc., their presence highlights the historical human values of the region. Around the centuries-old Victoria Garden, we rebuilt the courtyard to rehabilitate the neighborhood, and coordinated the architectural style as a method to carry out planning and architectural design for five blocks.

The maximum height of the historic block is 45m, the first and second floors are 12m high, and the top and roof are 10m high. The three-section facade is designed with tradition and balance. The west, north and east sides ensure the line-up rate of the building. The structure of the building is used to strengthen the connection with the street. On the east and west sides, the city roads are left to enter and exit, so that the inner courtyard and the street are connected and become an important part of the urban public space. The Ritz-Carlton's unique understanding of quality and the atmosphere of the historic district allow us to establish a "back home feeling for our customers", designed, operated under the guidance of tens of thousands of words "Working Standards". Experience the repeated comparison of the foyer, banquet hall, guest room, leisure and fitness, etc., to provide customers with the best experience with the most suitable scale and the most comfortable streamline. In order to maintain the harmony of the outer edge texture of the same cultural relics building, using real brick masonry, combined with traditional crafts, we specially draw three-dimensional large-scale detailed drawings of stone, block, doors and windows and roof to guide the on-site construction, now it seems that this is guaranteed completion of degree.

Since its opening in 2015, the Ritz-Carlton has become a flagship presence in the Marriott Group's Asian Historic District, bringing them great benefits and reputation. Its success also encourages us to focus on the scale, form and function while paying more attention to people's experience and atmosphere creation in the face of historical block renewal. We have strengthened our concept of rebuilding in the new urban culture. And the combination of traditional craftsmanship continues the historical heritage, providing the best place and realistic background for continuing to tell Tianjin's own story.

维多利亚公园一侧 | *picture beside the Victory Park*

拆除前杂乱无序 | before demolition: disorganized space

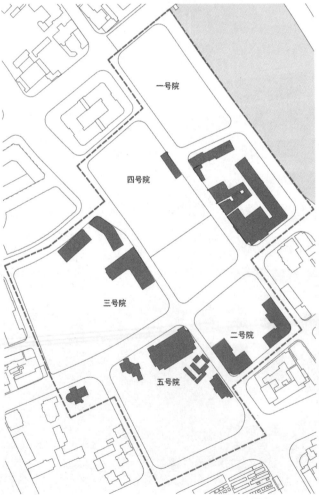

保留建筑 | save building

拆除后绿地空旷 | after demolition: green open space

建成后重塑都市空间 | after completion: reconstruction of urban space

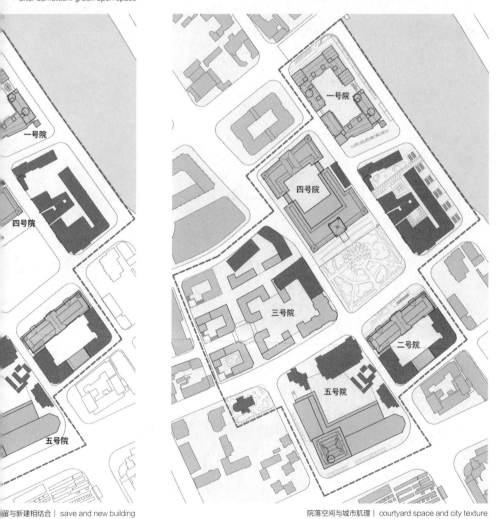

留与新建相结合 | save and new building

院落空间与城市肌理 | courtyard space and city texture

31

泰安道五大院鸟瞰图 | aerial view of five courtyards

烟台路两侧效果 | scene along Yantai Road

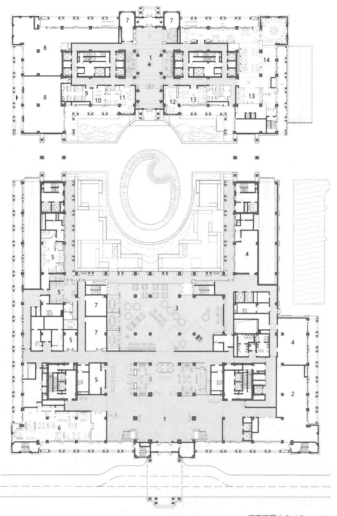

一层平面图 | first floor plan

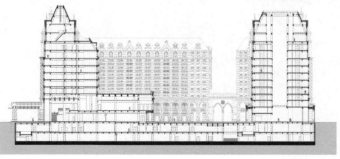

剖面图 | section

酒店大堂 | hotel lobby

二层走廊 | aisle in second floor

公寓大堂 | apartment lobby

37

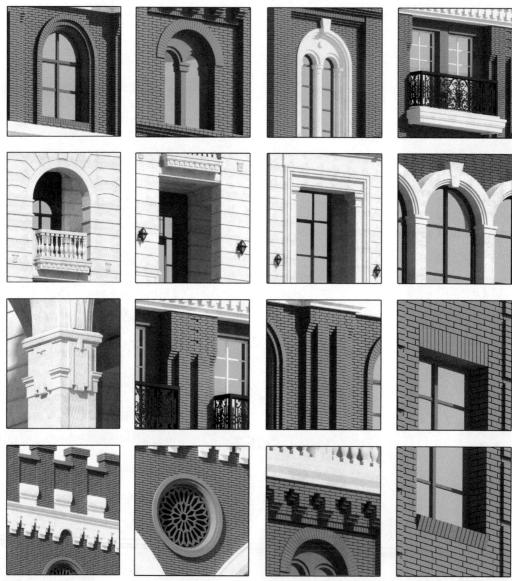

外檐细部 | detail of eave

大沽北路一侧雪景 | snow scene along North Dagu Road

酒店夜景 | night view of hotel

内院过廊下效果 | inner courtyard under the corridor

内院 | inner courtyard

大宴会厅单元 | grand banquet hall

大堂吧 | lobby bar

行政套房 | executive suite

宴会厅前厅 | banquet hall foyer

全日制餐厅装修 | decoration of all-day dining restaurant

decoration of all-day dining restaurant

中餐厅装修 | decoration of Chinese restaurant

意大利建筑师阿尔多·罗西说过："城市是生活在此地市民本身的集体记忆，城市本身是记忆的对象，又是承载记忆的场所，城市与记忆一样与物体和场所相连。"天津这座城市自建卫以来就被注入港口城市的基因，尤其是鸦片战争后成为多元文化的交汇区，变为西方建筑的荟萃之地。

我们在现实中往往欣赏由于时间推移而失去功能的元素，这些建筑的价值体现在其形式之中，它是一个常量。最大的谬误是只把经久建筑当作一个历史时代的作品，其实形式本身所蕴含的不同价值、意义和作用的能量是超越时代成为一类记忆的标志。在接到天津职业大学的设计任务后，我们通览了世界各大名校的建筑，项目组结合天津"万国建筑博览会"的城市特色，在充分满足职业教育复杂功能之后，利用当地红砖结合哥特风格提出反映教育类建筑特征的方案，得到校方认可。

在校园中轴线上布置一座高 60 m 的塔楼，彰显校园建筑特色和空间价值。主楼采用体量前后错动，在形

天津职业大学
Tianjin Vocational Institute

用地面积	262 537 m²
建筑规模	74 630 m²
建筑团队	天津市城市规划设计研究总院
获奖情况	2016 年度"海河杯"天津市优秀勘察设计建筑工程公建一等奖

外立面图 | facade

区位图 | location

体交接处插入特异体块的做法，打破其水平方向过长形成的呆板形象。同时，在端部结合角楼单元用竖向线条强化体量感，营造厚重形象。垂直方向采取三段式的处理方法，底层设拱券柱廊，中央二、三层有拱窗装饰，顶层山墙布置窄窗强调竖向的挺拔。两侧教学楼通过虚实结合的柱廊围合塑造院落空间，形成书院式教学单元。

西方哥特式建筑类型代表了大众对学院建筑形象的记忆，尽管哥特式与原始功能已经分离，但我们在结合用地、功能的条件下重新组合各元素符号，创造性地研究推敲各元素的结合方式和比例关系，使与形式融为一体的对学院形式的记忆重新焕发活力并赋予其意义，超越原始功能曾经定义的形式意义。民众记忆中的历史形式成为本次设计创新的催化剂和自主研发的基础。

Aldo Rossi once said, "The city is the collective memory of the citizens living here. The city itself is the object of memory, and it is the place to carry memories. The city is connected with objects and places like memories." Since the establishment of the city, the genes that have been injected into the port city, especially after the Opium War, have become the intersection of multiculturalism and become a gathering place for western architectures. It is not only a valuable architectural wealth for Tianjin western architectures, but also a memory of Tianjin's urban memory and an important carrier of history and culture.

In reality, we often appreciate the elements that lose their function over time. The value of these buildings is reflected in their form, which is constant. The biggest fallacy is that only the long-lasting architecture is a work of a historical era. In fact, the energy of different values, meanings and functions contained in the form itself transcends the times and becomes a kind of memory. After receiving the design task of the Architecture of Tianjin Vocational Institute, we have visited some of the world's major universities. The project team combined with the urban characteristics of the Tianjin, after fully understanding the complex functions of vocational education, using local red bricks combined with Gothic style to create a reflection. The program of educational building features is recognized by the school.

A 60m-high tower is placed on the central axis of the campus to highlight the architectural features and space value of the campus. The main building adopts the practice of inserting the monolithic body block at the intersection of the body, and breaks the rigid image formed by the long horizontal direction. At the same time, the corner unit is combined with the vertical line to strengthen the body mass to create a thick image. The three-stage treatment method is adopted in the vertical direction. The bottom arch vault colonnade has arch windows on the second and third floors, and the top gables are arranged with narrow windows to emphasize vertical straightness. The teaching buildings on both sides form a college-style teaching unit by enclosing the courtyard space through a combination of virtual and solid colonnades.

The Western Gothic architecture type represents the public's memory of the architectural image of the college. Although the Gothic form has been separated from the original function, we recombined the symbols of each element under the conditions of land use and function, and creatively studied the combination of elements and the proportional relationship makes the integration of the form and the meaning of the re-energization of the memory of the college form, beyond and deepens the formal meaning of the original definition of the original function. The historical form of the general memory of the masses becomes the catalyst and independent research and development of this design innovation.

鸟瞰图 | aerial view

英式建筑发展脉络　British architecture development process

16 世纪上半叶　　　　　　　　　　18 世纪下半叶到 19 世纪中叶　　　　　18 世纪下半

都铎风格　　　　　　　　　　　　　古典复兴

塔楼、雉堞、烟囱，体形多凹凸起伏，窗子的排列也很随意。结构、门、壁炉、装饰等常用平平的四圆心券，窗口则大多是方额的。在荷兰的影响下，常用红砖建造，砌体的灰缝很厚。腰线、券脚、过梁、压顶、窗台等则用灰白色的石材，很简洁。

人们开始攻击巴洛克与洛可可风格的烦琐、矫揉造作，极力推崇古希腊、古罗马艺术的合理性。

封建贵族一天一天地落，他们抚今追昔，起了美化中世纪的生

汉普顿宫 | Hampton Court Palace

大英博物馆 | British Museum

封蒂尔

剑桥大学 | University of Cambridge

爱丁堡大学 | The University of Edinburgh

19 世纪末期　　　　　19 世纪 30 年代到 70 年代　　　　　19 世纪下半叶到 20 世纪初期

哥特复兴　　　　　　　　折中主义

化的哀歌（模仿哥堂）。

在反拿破仑的战争中，欧洲各国的民族意识高涨，热衷于弘扬民族的文化传统。也反映了资本主义社会早期，一些人渴望自由，希望摆脱学院派古典主义因循守旧的作风。

折中主义任意选择与模仿历史上的各种风格，把它们组合成各种式样，又称为"集仿主义"。折中主义建筑并没有固定的风格，建筑语言混杂，但讲究对比例的权衡推敲，常沉醉于对"纯形式"美的追求。

Monastery of Sentier

英国住宅 | UK Housing

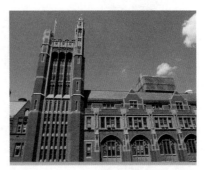

哥伦比亚大学教育学院 | School of Education, Columbia University

英国曼彻斯特建筑 | architecture in Manchester

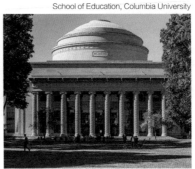

美国麻省理工学院 | Massachusetts Institute of Technology

大厦 | Eden Building

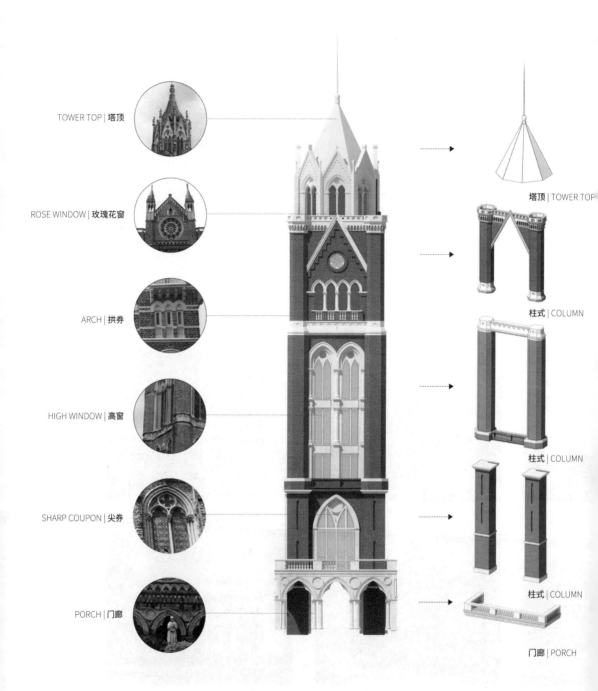

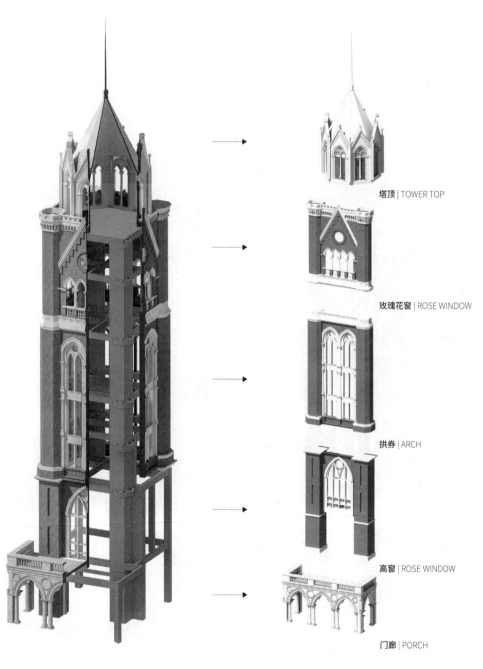

塔顶 | TOWER TOP

玫瑰花窗 | ROSE WINDOW

拱券 | ARCH

高窗 | ROSE WINDOW

门廊 | PORCH

语汇拆解 | architecture language disassembly

塔楼 | tower

立面图 | facade

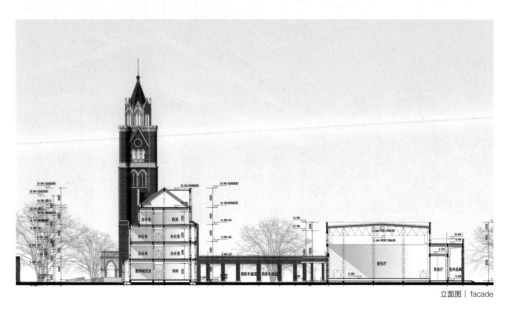

立面图 | facade

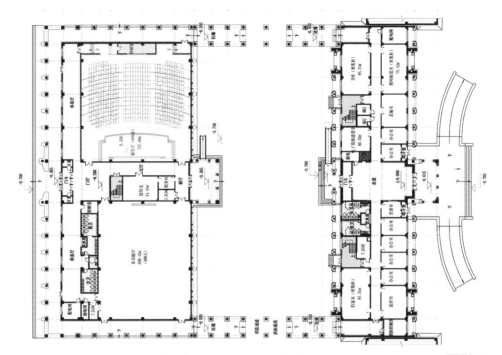

平面图 | plan

内院 2 | courtyard 2

内院 1 | courtyard 1

内院 3 | courtyard 3

券廊 | voucher gallery

清水混凝土 | architectural concrete

行政楼 | administration building

券廊 | voucher gallery

原天津市民族文化宫是全国第二座为各族人民文化事业提供专门活动场所的建筑，为了支持城市发展和老旧区改造，原民族文化宫于2006年被拆除并准备异地还建。用地选址几经变化，最终被调整到南运河与三条石大街交角处的一块三角形地块上，在政府兑现10年前承诺的背景下，我们开展了极限条件下的设计工作。

规划用地紧邻大运河保护区天津段的南运河，按照运河保护规划及三边管理规定，建筑退线不小于25 m，建筑高度与建筑投影至河岸距离的比例不能大于1∶1；沿三条石大街有一条规划的地铁四号线，其线位侵入用地红线，按规定地铁控制线内严禁建设，这限制了地上及地下空间的开发利用；基地西侧与建成的居民区相连，建筑必须退让足够距离。这些限制就像无形的绳索将用地及地上、地下建筑紧紧绑在一起，在苛刻的条件下我们重新梳理用地内各种流线，发现建筑出入口位置、地下车库出入口位置、建筑高度以及建筑面积都没有可供选择的余地，严重压缩了建筑师的创作空间。

天津市民族文化宫
Tianjin Culture Palace of Nationalities

总用地面积｜10 724 m²
可用地面积｜6982 m²
总建筑面积｜17 200 m²
建筑团队｜天津市城市规划设计研究总院
内装设计｜福建永盛设计装饰工程有限公司
策展设计团队｜北京正邦兴业建筑技术开发有限公司
景观设计团队｜浩安生态环境建设有限公司
剧场顾问｜天津北方文化科技有限公司
获奖情况｜2021年度"海河杯"天津市优秀勘察设计建筑工程公建三等奖

效果图｜effect picture

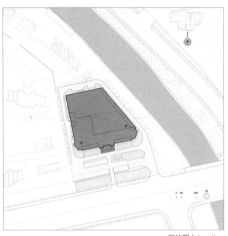

区位图｜location

在不到11 000 m²规划用地上，满足条件的可建设用地不足7000 m²，但在此可用地上需建设约17 000 m²的建筑才能满足业主要求。任务书要求民族文化宫要有小剧场（500个座位）、民族手工展示区、民族文艺培训区、民族体育训练区、篮球场以及办公区、管理区、食堂等。同时剧场、篮球场还要保证单独使用。针对功能流线复杂、空间利用强度高的特点，我们采用空间规划和体块模拟分析的方法，叠加各种空间组成有机整体，在多种可能的方案中选择空间效果好和利用效率最高的进行深化。立面选择红砖作为展现地方特色的材料，采用完全符合各种限制要求的体形来恰如其分地表达用地上看不到的真实，创造性地传承了原民族文化宫的风格。

建筑师在面对来自场地的苛刻条件、来自项目的极限需求和来自业主的审美要求时，需要高超的专业技能、耐心和沟通技巧才能平衡各种诉求。民族文化宫的方案就是在这种复杂与矛盾的现实中，找到了释放基地极大潜力，以及展现当代技术条件下地域建筑魅力的解决方案，这是我们经历民族文化宫设计的最大收获。

The former Tianjin Culture Palace of Nationalities is the second building in China to provide special activities for the cultural undertakings of all ethnic groups. In order to support urban development and renovation of old areas, the Cultural Palace was demolished in 2006 and is ready to be built in other places. After several changes in the site selection, it was finally adjusted to a triangular ground at the corner of the South Canal and Santiaoshi Street. In the context of the government's commitment to fulfill more than 10 years ago, we started the design work under extreme conditions.

The planned land is adjacent to the South Canal of the Tianjin section of the Grand Canal Protection Zone. According to the canal protection plan and the trilateral management regulations, the building exit line is not less than 25m, and the ratio of the height of the building to the distance from the building to the river bank cannot be greater than 1 : 1; Along Santiaoshi Street, there is a planned subway Line 4, which invades the red line of the land. Construction is strictly prohibited in the subway control line. This limits the development and utilization of the above-ground and underground space. The west side of the base is connected to the built-up residential area, and the building must be retired distance. These restrictions are like invisible ropes that tie the land to the ground and underground buildings. Under the harsh conditions, we re-combine the various streamlines in the land and find the entrance and exit of the building, the entrance and exit of the underground garage, the height of the building and the building area. There is no room for choice, which seriously reduces the space for architects to create.

On the nearly 1.1 hectares of planned land, only 0.7 hectares of land can be used to meet the conditions, but it is necessary to build 17,000 square meters of buildings on the available land to meet the requirements of the owners. The design task requires that the National Culture Palace should have the following spaces: a small theater (500 seats), an ethnic hand-painted exhibition area, national literary and artistic training, national sports training, basketball courts, office, management, canteens, etc. At the same time, the theater and basketball court must be used separately. In view of the complex function streamline and high space utilization intensity, we use spatial planning and block simulation analysis methods to superimpose various spaces to form an organic building,and select the space effect and the most efficient use among various possible solutions. The facade selects red brick as a material to show local characteristics, and adopts a body shape that is completely unified with various restrictions to properly express the realities that cannot be seen on the land, and creatively inherits the style of the original National Culture Palace.

Architects face the harsh conditions from the venue, from the extreme needs of the project, from the owner's aesthetic tendency, the need for professional skills, technical patience and communication skills to balance the various demands. The program of the National Culture Palace is to find a solution to the great potential of the release base and to show the charm of regional architecture under the conditions of contemporary technology in this complex and contradictory reality. This is the greatest achievement we have experienced in the design of the Tianjin Culture Palace of Nationalities.

实景 | scene

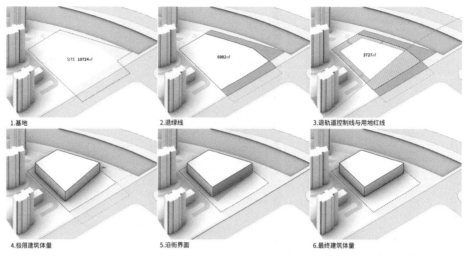

1.基地　　2.退绿线　　3.退轨道控制线与用地红线
4.极限建筑体量　　5.沿街界面　　6.最终建筑体量

场地限制条件 | site limitation

施工现场 | under construction

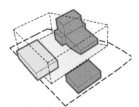

不宜分散布置 |
not suitable for scattered layout

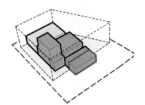

不宜同层集中布置 |
not suitable on the same floor

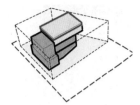

不宜叠放布置 |
not suitable for folded

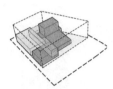

紧密结合附属用房 |
combined with auxiliary room

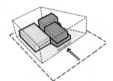

前厅空间局促 |
narrow lobby space

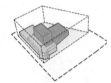

剧场及附属用房 |
theatre and auxiliary room

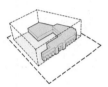

公共服务 |
public service space

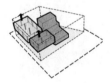

无柱空间上移 |
move up the no n-pillar space

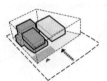

前厅偏置 |
offset the lobby

球类馆及其他运动馆 |
gymnasium

多功能厅及艺术培训等 |
multi-function hall

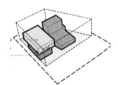

避免动静交叉 |
separating dynamic zone from static zone

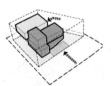

独立出入口 |
separated entrance

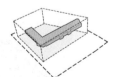

业务用房 |
business room

功能组织 |
function organazion

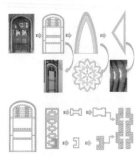

元素提取 | extraction element

窗户式样 | window style

门厅式样 | hall style

主入口处实景图 | main entrance

改造前 | before renovation

改造前 | before renovation

改造前 | before renovation

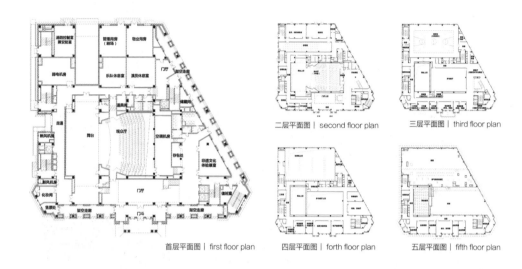

首层平面图 | first floor plan
二层平面图 | second floor plan
三层平面图 | third floor plan
四层平面图 | forth floor plan
五层平面图 | fifth floor plan

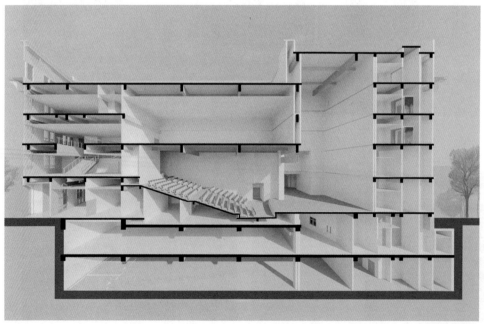

剖面图 | section

室内透视图 | interior perspective view

室内透视图 | interior perspective view

实景 | scene

实景 | scene

走廊 | corridor

投标方案二 | bidding plan

场地 | site

分割 | separated

抬升 | lift up

效果图 | effect picture

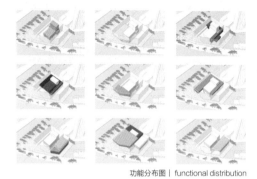

功能分布图 | functional distribution

鸟瞰图 | aerial view

外立面图 | facade

投标方案三 | bidding plan

在寻求民族性的表达途径中，我们为重建的民族文化宫寻求三个突破点——对称、层叠、向上。同时，我们在建筑风格上，利用有力度的建筑体量，具有民族特色的建筑造型，以及具有民族符号纹理、窗框装饰等特色雕饰，充分体现了民族文化宫文化一家亲的融合和统一。

In seeking ways to express nationality, we seek three breakthrough points for the reconstructed Cultural Palace of Nationalities—symmetry, cascading, and upward. At the same time, in terms of architectural style, we use the powerful building volume, the architectural shape with national characteristics, and the characteristic carvings with national symbol texture, window frame decoration, etc., which fully reflects the fusion and unity of the cultural of the Cultural Palace of Nationalities.

鸟瞰图 | aerial view

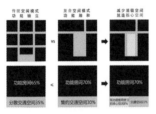

空间分析图 | space analysis

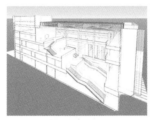

位置分析图 | site analysis

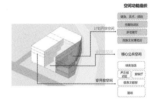

功能分析图 | function analysis

外立面图 | facade

中庭 | atrium

效果图 | effect picture

国际上把 60 岁以上人口占总人口比重达到 10%，或 65 岁以上人口占总人口的比重达到 7% 作为国家或地区进入老龄化社会的标准。2017 年我国 60 岁以上老龄人口达到 2.4 亿，占比突破 17%，中国未富先老的社会结构已经到来，但居家养老社区、适老住宅的建设却相对滞后，远不能适应现实的巨大需求。五福里的建设就是在城市更新的背景下结合混合社区开放式住区的理念，为适老住宅的设计实践提供难得的机会。

当前中国住宅的建设几乎都被低市场风险的公式化手法彻底统治，尤其是商品房，以日照、停车、绿化率为导向的追求利益最大化的结果就是毫无生机的兵营式小区。但在五福里，我们尝试从规划入手，在实现窄路密网活力街区的形态之外，强化配套的数量和质量，建立社区日间照料中心、医养结合的社区医院等提供专业化服务的机构，使居家养老成为可能。

适老住宅的设计是从住宅入口开始的，入口坡道的坡度不能超过 20%；信报箱的底边距地面不小于

黑牛城道五福里

Wufu Community, Heiniucheng Road

用地面积	190 675.90 m²
建筑规模	428 100 m²
建筑团队	天津市城市规划设计研究总院建筑一院
合作团队	天津方标世纪规划建筑设计有限公司
获奖情况	2019年度中国土木工程詹天佑奖优秀住宅小区金奖
	2018年度"海河杯"天津市优秀勘察设计住宅与住宅小区一等奖

鸟瞰图 | aerial view

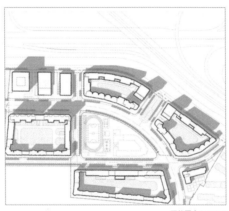

区位图 | location

600 mm，且高度不能超过1600 mm；入户门把手等不能用圆形，把手要能用手握住；入户门外设置户外手提袋挂钩，给老人用钥匙开锁提供方便；门厅除设置换鞋用的鞋凳之外，在旁边要设置便于起身的扶手；卫生间用推拉门，防止在浴室内滑倒后无法开门；坐便器、淋浴间要加装扶手，同时轮椅能回转；厨房设置"工作三角区"省时省力安全，水龙头采用拉拔式，操作台下部留出空间方便坐在轮椅上操作……这些对年轻人来说看似微不足道的细节，都是老人生活中可能遇到的困难，设计师认真思考进行改进完善并付诸实践，让老人们能感受到他们期待的、如同家人般的关怀和体贴。

适老住宅的建设让人们了解到有这么一种设计，不仅对老年人而且对青年人和中年人也给予温暖的关爱。这种带有生活预见性和生活智慧结晶的住宅比你想要的更实际，在设计之初就预留了关怀的可能，随着时光流逝逐渐变为现实，我们是时候来认真思考为父母、为将来的自己在适老住宅上做些什么了。

天津河西区

2014—2016

The international standard for the aging society is that the elderly population over 60 years old accounts for more than 10% or over 65 years old accounts for 7%. In 2017, China's aging population reached 240 million, accounting for 17% of the total. China is marching into the aging society, but the construction of home-age care communities and old-suitable houses is quite lagging far from being able to adapt to the huge demand of reality. The construction of Wufuli Community is to combine the concept of open residential areas with mixed communities in the context of urban renewal, providing appropriate design for the old population.

At present, the production of Chinese houses is almost completely ruled by the formula of low market risk, especially commercial houses. The pursuit of maximum benefit by sunshine, parking, and greening rate is the lifeless barracks. However, in Wufuli Community, the designers tried to start from the planning, to achieve the quantity and quality of the supporting facilities outside the narrow road network, and to establish a community day care center, a community hospital that combines medical care and other professional services. Home care is possible.

The design of the house for senior citizen starts from the entrance of the household. The slope of the entrance ramp cannot exceed 20%; the bottom edge of the letter box is not less than 600mm from the ground, and the height cannot exceed 1600mm; the door handle cannot be round handle. It is necessary to hold it by hand; set an outdoor tote bag hook outside the entrance door to provide convenience for the elderly to unlock with a key; in addition to setting up a shoe bench for changing shoes, the hall should be provided with a handrail for easy lifting; a sliding door for the bathroom prevent the door from slipping in the bathroom and open the door; the toilet and shower should be equipped with handrails, and the wheelchair can be rotated; the kitchen setting "work triangle" saves time and effort, the faucet adopts pull-out type, and the lower part of the console leaves space for the people sitting in the wheelchair. These seemingly insignificant details for young people are all difficulties that may be encountered in the life of the elderly. The designers seriously think about improving and perfecting and putting them into practice so that the aged can feel what they expect from family-like care and thoughtfulness.

The construction of houses for the aged allows people to understand that there is such a design that not only provides warmth and care for the elderly but also the young and middle-aged. This kind of home with life expectation and wisdom of life is more practical than you want. At the beginning of the design, the possibility of caring is reserved. As time goes by, it becomes a reality. It is time for us to seriously think about it for parents, and for the future.

道路一侧实景图 | scene photo along the road

立面图 | facade

鸟瞰图 | aerial view

立面图 | facade

85

实景 | scene

实景 | scene

实景 | scene

① 单行辅路断面

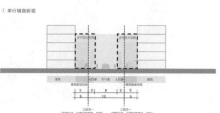

② 生活性道路断面(两侧均为建筑)

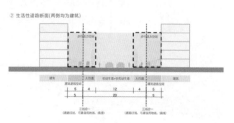

道路剖面图 | road section

实景 | scene

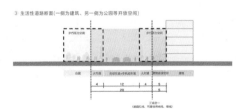

③ 生活性道路断面（一侧为建筑、另一侧为公园等开放空间）

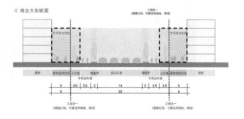

④ 商业大街断面

道路剖面图 | road section

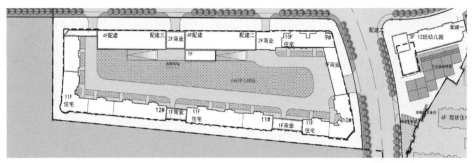

院落平面图 | courtyard plan

院落实景 | courtyard

院落实景 | courtyard

院落实景 | courtyard

配建（菜市场、社区医疗服务中心）| Supporting building

幼儿园 | kindergarden

沿街外景 | scene along the street

小学 | primary school

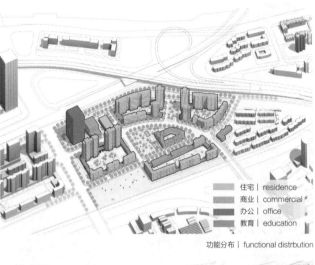

住宅 | residence
商业 | commercial
办公 | office
教育 | education

功能分布 | functional distrbution

路网分析 | analysis of road network

院落入口标识 | sign of courtyard entrance

大堂标识 | sign of lobby

入户门标识 | sign of entrance door

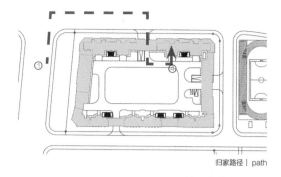

归家路径 | path

沿街标识 | road signs

95

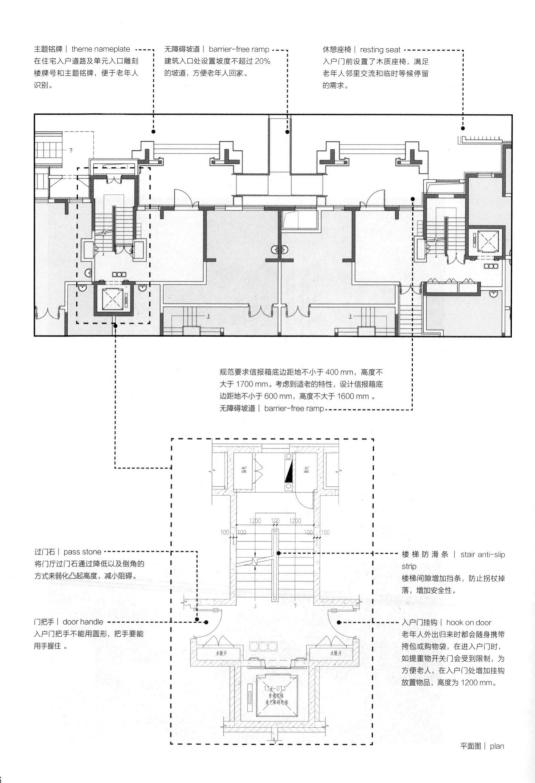

平面图 | plan

入户大堂 | lobby

无障碍坡道 | barrier-free ramp

信报箱 | letter box

入户门挂钩 | hook beside door

门把手 | door handle

过门石 | pass stone

楼梯防滑条 | stair anti-slip strip

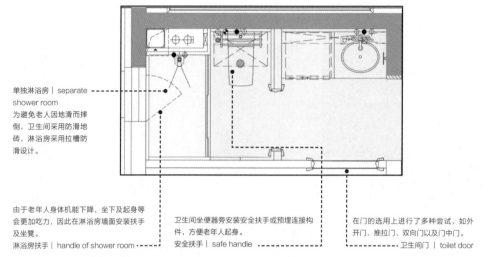

单独淋浴房 | separate shower room
为避免老人因地滑而摔倒，卫生间采用防滑地砖，淋浴房采用拉槽防滑设计。

由于老年人身体机能下降，坐下及起身等会更加吃力，因此在淋浴房墙面安装扶手及坐凳。
淋浴房扶手 | handle of shower room

卫生间坐便器旁安装安全扶手或预埋连接构件，方便老年人起身。
安全扶手 | safe handle

在门的选用上进行了多种尝试，如外开门、推拉门、双向门以及门中门。
卫生间门 | toilet door

平面图 | plan

卫生间双向门 | two-way door of toilet

坐便器安全扶手/单独淋浴房 | toilet safe handle /seperated shower room　　淋浴房扶手 | shower room handle

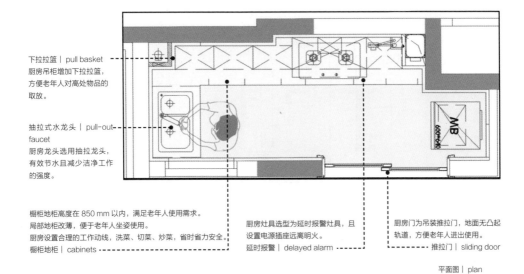

下拉拉篮 | pull basket
厨房吊柜增加下拉拉篮，方便老年人对高处物品的取放。

抽拉式水龙头 | pull-out faucet
厨房龙头选用抽拉龙头，有效节水且减少洁净工作的强度。

橱柜地柜高度在 850 mm 以内，满足老年人使用需求。
局部地柜改薄，便于老年人坐姿使用。
厨房设置合理的工作动线，洗菜、切菜、炒菜，省时省力安全。
橱柜地柜 | cabinets

厨房灶具选型为延时报警灶具，且设置电源插座远离明火。
延时报警 | delayed alarm

厨房门为吊装推拉门，地面无凸起轨道，方便老年人进出使用。
推拉门 | sliding door

平面图 | plan

抽拉式水龙头 | pull-out faucet

延时报警 | delayed alarm

橱柜 | cupboard

下拉拉篮 | pull down basket

橱柜地柜 | cabinets

墙角处理 | corner treatment
墙体阳角处以圆角的方式处理，降低磕碰对老人的伤害。

辅助照明筒灯 | auxiliary lighting
考虑老年人的灯光需求，尤其是在读书看报的情况下，老年人的灯光需求是年轻人的两倍，且部分辅助光源单独设置开关控制，节能减耗。

折叠换鞋凳 | folded shoes changing tool
入户玄关柜高度 850 mm，可做支撑台面；应配置坐凳，方便坐姿换鞋；坐凳高度控制为 300 mm，坐面不宜过深。

在卧室及通往卫生间的过道中设置小夜灯，为老年人夜间走动提供方便。
走廊小夜灯 | corridor night light

户内过门石仅高出地面 3 mm，且设置斜角倒坡，无明显脚感，方便老年人进出。
过门石 | pass stone

出于老年人安全考虑，设置紧急报警按钮，距地面 650mm，满足老年人使用需求。
紧急报警按钮 | panic button

为满足老年人对晾洗和储物的需求，在客厅阳台位置增加晾衣杆和储物柜。
阳台及储物柜 | balcony&locker

平面图 | plan

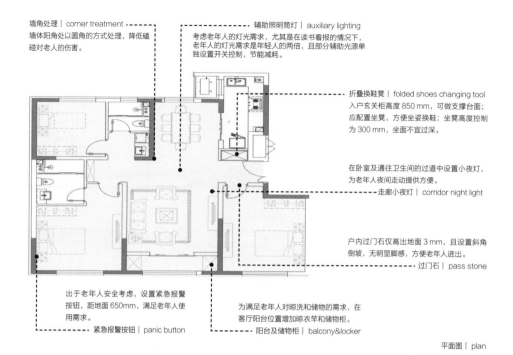

销售样板间 | sample room

辅助照明筒灯 | auxiliary lighting

紧急报警按钮 | panic button

阳台及储物柜 | balcony & locker

折叠换鞋凳 | folded shoes changing tool

走廊小夜灯 | corridor night light

墙角处理 | corner treatment

过门石 | pass stone

天津文化中心建设接近尾声时，周边地块的开发建设被提上议程。作为还在规划中的未来城市中心有相当的不确定性，为了不被未知区域淹没，建筑需要结合环境成为引领地域发展的标志。我们与建筑师里查德·安格里斯（Richard Angelis）一起投标，因采用先进理念及场地利用合理获得天津市第四中学迁址扩建工程的设计权。

新校舍规划用地 89 770 m^2，可用地 53 000 m^2，建筑面积 57 638 m^2，预计可容纳 80 个班，供 4500 名师生使用。按现行规范人均用地与建设指标均不满足要求，再加上用地东、南两侧濒临城市干道，东南角地上有高架桥，西北角地下有地铁线。外部近乎苛刻的条件和校方对开放式教学的追求让我们不得不摒弃传统设计方法，采用"总体布局集约、公共空间复合、建筑景观一体"的策略，将操场布置在西北角地铁线上，用长廊统一各教学空间及生活用房形成整体形态来平衡大尺度高架桥带来的压力。

设计真正的挑战来自与校方进行教学理念的沟通。学校是培养创造力的场所，应最大限度地促进师生、学生

天津市第四中学迁址扩建工程
Tianjin No.4 Middle School Relocation and Expansion Project

用地面积 ｜ 89 770 m²
建筑规模 ｜ 57 638 m²
建筑团队 ｜ 德阁建筑设计咨询（北京）有限公司、天津市城市规划设计研究总院
获奖情况 ｜ 2021年度海河杯天津市优秀勘察设计建筑工程公建一等奖

鸟瞰图 ｜ aerial view

区位图 ｜ location

之间的交流，所以我们建议校方采用开放式教学方法，营造功能混合的学习场所。另外，学校教学应以学生为主体，自己决定并选择课程。校方最初以师资有限为由婉拒我们提出的开放式教学的方案，几经讨论部分保留了我们的想法。入口广场采用底层架空的形式结合内院布置，提高空间使用效率的同时营造了礼仪性的灰空间。首层布置大空间教学用房，如阶梯教室，二至五层布置实验室和教室，中央走廊成为各层学生课余交流的主要场所，教学楼底层与运动场看台结合形成实用性与舒适性兼备的交流场所，让学生在学习和生活中感受到交流带来的乐趣。

2018年天津所有中学实行选课制教学，学生按选择的专业自主选课，决定受教育的内容。正是由于我们的预想和实践，使四中校方能从容应对突如其来的变化，校方因此对我们的工作非常肯定并心怀感激。建设现代建筑的初心就是面对人类社会、经济发展带来的各种矛盾，不断创新提出合理且具有前瞻性的解决方案，用设计的智慧提高人们的生活质量和促进社会公平的实现。

When the construction of the Tianjin Cultural Center came to an end, the development and construction of the surrounding plots was put on the agenda. As a future urban center still under planning, there is considerable uncertainty. In order not to be inundated by unknown areas, the building needs to be combined with the environment to become a symbol of regional development. We bid together with architect Richard Angelis for the right to design with advanced concepts and venues.

The new school building has land area of 89,770 square meters, available land area of 53,000 square meters and building area of 57,638 square meters. It is expected to accommodate 80 classes for 4,500 teachers and students. According to the current regulations, the average land use and construction indicators are not satisfactory. In addition, the east and south sides of the land are on the side of the main road, the southeast corner has a viaduct, and the northwest corner has a subway line. The harsh external conditions and the pursuit of open teaching make us have to abandon the traditional design method, set the playground on the northwest corner subway line, and use the corridor to unify the teaching space and the living room to form a whole form to balance the large scale. The pressure brought by the viaduct.

The real challenge of design comes from communicating with the school's teaching philosophy. The school is a place to cultivate creativity. Its greatest potential comes from promoting communication between teachers, students and students. Therefore, we suggest that the school adopt an open teaching method to create a mixed learning place. In addition, the school offers academic course based on students' interests and development . The school initially refused our open teaching program on the grounds of limited teachers, and some of the discussions retained our ideas. The entrance plaza adopts the underground overhead form combined with the inner courtyard layout to improve the space use efficiency while creating a ceremonial gray space. The first floor is equipped with a large space teaching room, such as a lecture hall. Laboratories and classrooms are arranged from the second to the fifth floor. The central corridor becomes the main place for students to play and talk after class. The bottom of the teaching building and the stadium stands combine to form a practical and comfortable exchange place, so that students can enjoy the fun of communication in their study and life.

In 2018, all secondary schools in Tianjin implemented elective courses teaching, so the students can choose their own courses according to their chosen majors to determine the content of education. It is precisely because of our advanced expectations and practice that the No. 4 Middle school can calmly respond to sudden top-down changes, and the school is therefore very positive and grateful for our work. The initial intention of modern architecture is to face various contradictions brought about by human society and economic development. Constantly innovate and propose reasonable and forward-looking solutions, and use design wisdom to improve people's quality of life and lead social fairness.

实景 | scene

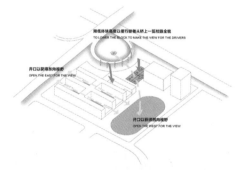

场域分析图 | field analysis

鸟瞰图 | aerial view

一、总体布局集约化 | Intensive Overall Layout

在平面的设计上，坚持总体布局集约化的理念，将学校分为教学区、生活区和运动区三部分。教学区引入"教学综合体"的概念。

In the plan design, adhering to the principle of intensive overall layout, the school is divided into three areas: teaching area, living area and sports area Points. The teaching area introduces the concept of "teaching complex".

分区功能布局 | Function

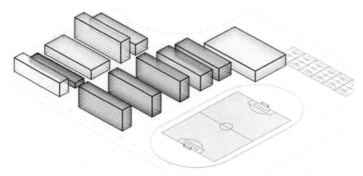

传统布局略显零散，各功能之间没有交流，不利于学生综合全面发展。另外空间使用率低，交通空间过多，使得本就用地紧张的地块更加拥挤。
The traditional layout is slightly scattered, thus there is no communication between different functions. It is not conducive to the comprehensive development of students. Moreover, low space utilization rate and large traffic space make the site more crowded.

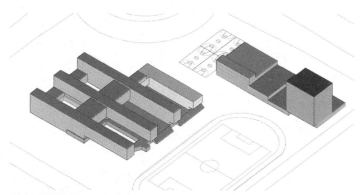

混合布局能有效地增强各功能之间的联系，并最大化节约土地，提高单位面积的使用率。
Hybrid layout can effectively improve the connection between functional areas, maximize land savings, and improve the usage of unit area.

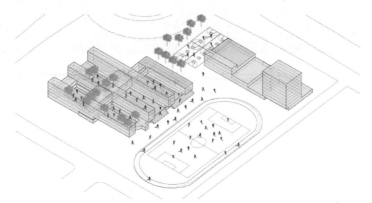

在校园绿化景观设计上，提出了"建筑景观一体化"的设计理念，引入了景观性看台和特色院落。
In the campus green landscape design, the architects put forward the "integrated architecture and landscape" design concept, introducing landscape grandstand and featured courtyard.

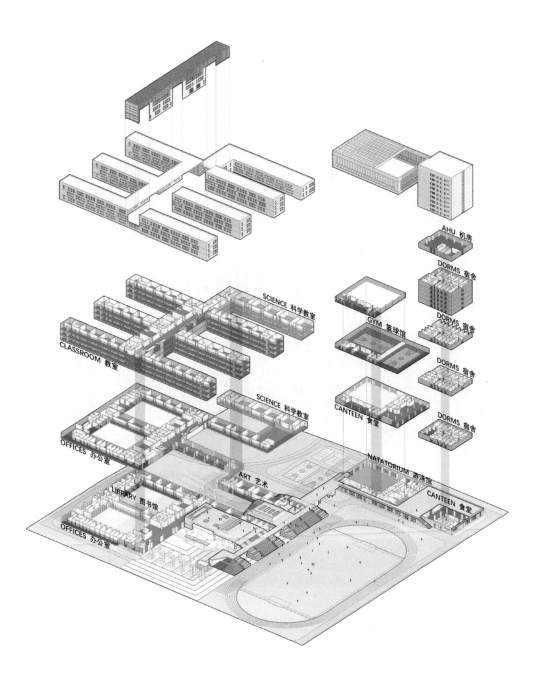

爆炸图 | exploded diagram

二、建筑景观一体化 | Integrated architecture and landscape

景观性看台 | landscape grandstand

这一设计理念改变传统看台配套附属的消极作用，与整个教学区建筑积极结合，与屋顶绿化连为一体，加强教学区与运动场的联系，打造形式多样、功能复合的景观性看台，最大限度地利用一切来美化环境。

实景 | scene

实景 | scene

实景 | scene

This concept changes the negative effect of the traditional grandstand supporting the annex, but actively connect the whole teaching area building, integrating with the roof greening. It strengthens the connection between the teaching area and the sports field, and forms a landscape grandstand with various functions.

延续院落格局 | Continue the existing courtyard pattern

延续四中老校区精致的院落格局和优美的绿化环境，综合体内穿插了四个院落空间，可以容纳老校区保留的小品、雕塑、绿植，体现了"习礼大树下"的中国传统建筑思想。

实景 | scene

实景 | scene

实景 | scene

The new campus continues the existing exquisite courtyard pattern and beautiful green environment of the old campus. The complex is interspersed with four courtyard spaces, which can accommodate the retained skits, sculptures and green plantings, reflecting the traditional Chinese architecture thought of "trained in ritual under the tree".

三、公共空间复合化 | complex public space

四轴体系 | Four-axis system

建筑通过"两横两纵"四个空间轴，形成棋盘网格状空间结构。这些交通空间作为充满活力的连接介质，将所有的教学功能在平面上连接成一个有机的整体。

The building forms a chessboard grid-shaped spatial structure through four spatial axes of "two horizontals and two verticals". These traffic spaces serve as a vibrant connecting medium, connecting all teaching functions on a flat surface into an organic whole.

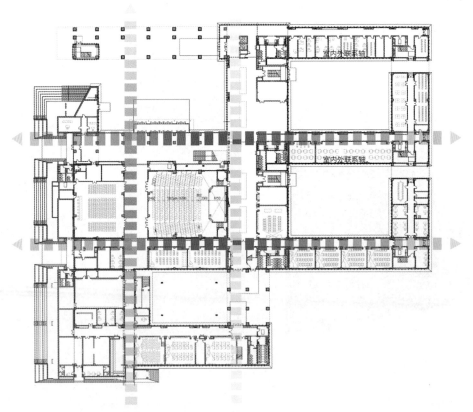

平面图 | plan

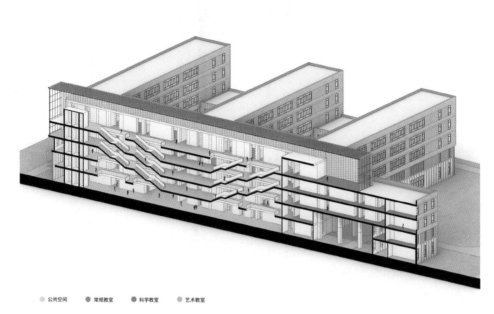

○ 公共空间　● 常规教室　● 科学教室　● 艺术教室

公共空间轴剖面图 | shaft profile of public space

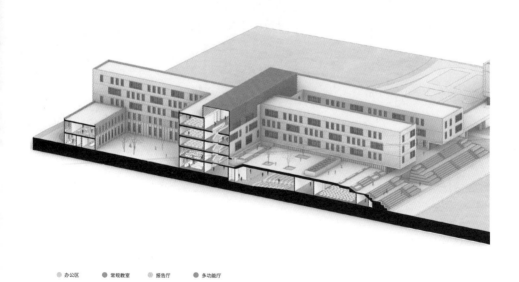

● 办公区　● 常规教室　● 报告厅　● 多功能厅

教学楼轴剖面图 | shaft profile of teaching building

在垂直空间上,通过教学轴的 T 形共享楼梯,又将五层平面紧紧串联在一起。

In vertical space, the five floors are connect tightly through the T-shaped shared stairs of the teaching axis.

T形楼梯 | T stairs

实景 | scene

教学楼立面图 | teaching building facade

实景 | scene

实景 | scene

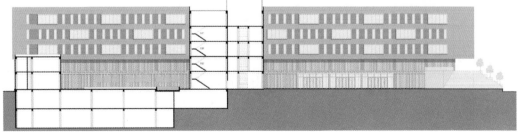

教学楼立面图 | teaching building facade

实景 | scene

左图：宿舍楼与体育馆南立面图 | south Dorm and Gym facade
中图：体育馆东立面图 | east Gym facade
右图：宿舍楼东立面图 | east Dorm facade

实景 | scene

实景 | scene

实景 | scene

实景 | scene

室内实景 | interior scene

模型图 | building model

模型图 | building model

在中国经济经过 40 多年的快速发展给城市带来了翻天覆地的变化之后，在物质需求得到满足之后，文化的振兴成为社会全员的共识和企盼。天津文化中心从决策到建设正是当时天津市政府对人民希冀的回应，也是展现市政府文化兴市雄心的惠民工程。

2008 年我们同株式会社山本理显设计工场共同投标天津图书馆工程并中标，承担了图书馆的建筑设计工作，同时还肩负起文化中心各个场馆方案整合以及修建性详细规划编制工作。

随着互联网的发展，信息时代使人们获取知识的渠道和手段日新月异，包括图书在内的传统纸媒逐渐式微并被新电子媒体所取代，发行了 200 多年的《大英百科全书》停止发行印刷版就预示着一个旧时代的结束和一个新纪元的开始。图书馆作为纸版图书收集、整理和发布的场所，其利用模式已经发生改变，读者对于自由交流、体验阅读和随机相遇的需求必将推动图书馆的公共空间向更开放、更融合和更包容的方向发展。我们将图书馆的设计目标定位为"空中读书乐园"。创造为读者带来新体验的知识乐园，成为设计师追寻的共同目标。

TianJing Library 天津图书馆

用地面积 | 36 676 m²
建筑规模 | 57 125 m²
建筑团队 | 株式会社山本理显设计工场、天津市城市规划设计研究总院
结构顾问 | 三江钢结构设计有限公司、天津大学
内装设计 | 株式会社山本理显设计工场、天津市城市规划设计研究总院、南洋装饰设计公司
智能化设计 | 天津中发机电工程有限公司
获奖情况 | 2015 年度全国优秀工程勘察设计行业奖一等奖
2014 年度"海河杯"天津市优秀勘察设计建筑装饰一等奖
2013 年度"海河杯"天津市优秀勘察设计建筑工程特别奖
2013 年度"海河杯"天津市优秀勘察设计专项工程(弱电工程)特别奖
2013 年度天津市钢结构金奖 – 优秀设计奖

外立面图 | facade

区位图 | location

依据读者的行为模拟对任务书重新整合分类,按公共性的强弱将能够开放的功能空间全部开敞,首层布置问询处、儿童及老年阅览、报告厅,二层、三层为一般阅览及期刊阅览,四层为电子阅览、音像阅览,五层布置古籍整理、修复以及研究用房。功能布局体现由动到静、由开放到封闭、由流动到固定的趋势。结合山本理显的空间理念,结构师创造性地利用"钢框架支撑旋转交错桁架"体系,实现了空间形式的自由流动,摆脱了重力带来的约束,释放了空间的活力,为空间形式的自由提供了技术保障。

"钢框架支撑旋转交错桁架"体系荣获 2013 年度天津市科技进步一等奖,这是对其技术先进性的表彰,但其意义不止于此,意味着人们对空间的感知和体验因这项技术进步向前跨越了一大步。图书馆空间成为在现行技术条件和规范下创造自由流动空间的一种范式,更是人类追求建筑空间自由的阶梯。

外表朴素、内部空间体验丰富的图书馆表达了世界越来越轻量化且呈均质状态的空间意象,新型结构使空间摆脱重力束缚,描绘出无限可能,空间形式蕴藏着对未来生活的预告,也折射出天津由传统城市向现代化都市转变的身影。

After more than 30 years of rapid development, the economy of China has brought profound changes to the city. After the material shortage is met, the revitalization of culture has become the consensus and hope of all citizens. From the decision-making to the construction of the Tianjin Cultural Center, it was the response of the Tianjin Municipal Government to the people's hopes at that time, and it was also the people-benefit project that showed the ambition of the city government culture.

In 2008, Architecture branch institute of Tianjin Urban Planning and Design Institute jointly bid for Tianjin Library with Riken Yamamoto & Field Shop, and responsible for the architectural design work of the library. At the same time, Architecture branch institute also hold an important position in integrating of various design groups and the construction of detailed planning.

With the development of the Internet, the information age has made people's access to knowledge change rapidly. New electronic media has gradually replaced the traditional paper media, including books. The publication of the *Encyclopedia Britannica* for more than 200 years indicates the end of an old era and the beginning of a new era. As a place for collecting, organizing and publishing paper-based books, the use mode of the library has changed. The readers' need for free communication, experiencing reading and random encounter will promote the library's public space to be more open, more integrated and more inclusive. We have set the library's design goals as a "knowledge paradise." Creating a knowledge paradise that brings new experiences to readers has become a common value pursued by designers.

According to the reader's behavior simulation, the design task book will be re-integrated and classified, and the open functional space will be opened according to the public strength. The first floor will be arranged for enquiries, children, elderly reading and reporting halls. The second and third floors are for general. For reading and periodical reading, the fourth floor is for electronic reading and audiovisual reading, and the arrangement of fifth floor is ancient books, restoration and research. The functional layout reflects the trend from static to flow, from flow to fixed, from fixed to open. Combining the space concept of Riken Yamamoto, the architect creatively uses the "steel frame supported rotating staggered truss" system, realizes the free flow of space form, frees the binding force from gravity, releases the vitality of space, and provides space for free freedom and technical support.

The library with a simple appearance and rich internal space expresses the space intention of the world's increasingly lightweight and homogeneous state. The new structure makes the space free from gravity and depicts infinite possibilities. The spatial form contains the preview of future life. It also reflects the transformation of Tianjin from a traditional city to a modern city.

夜景实拍图 | scene photo of night view

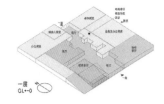
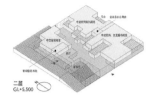

设计草图 | design sketch

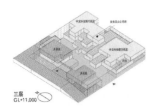
三层
GL+11.000

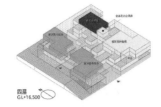
四层
GL+16.500

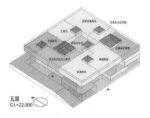
五层
GL+22.000

功能空间关系图 | spatial relationship

未来的画像 | future scene

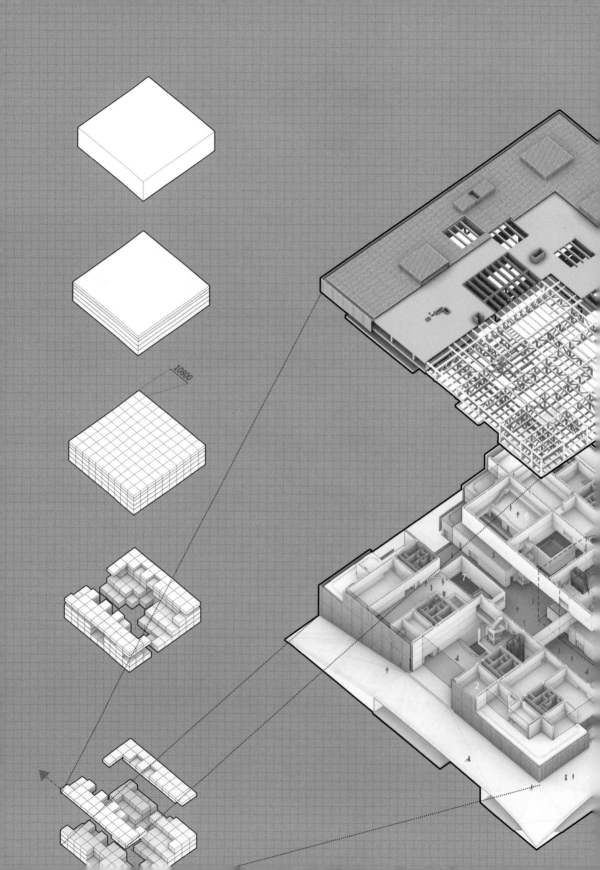

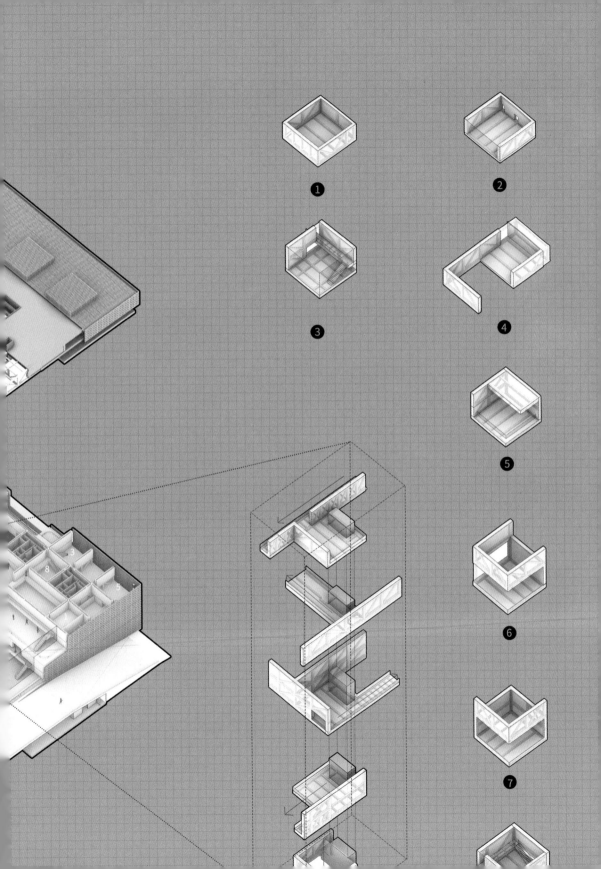

一层平面图 | first floor plan

三层平面图 | third floor plan

四层平面图 | forth floor plan

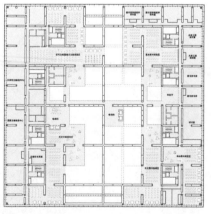

五层平面图 | fifth floor plan

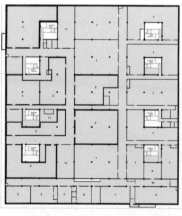

地下一层平面图 | basement floor plan

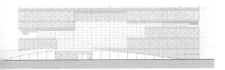

南立面图 | south facade

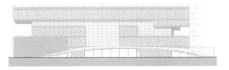

西立面图 | west facade

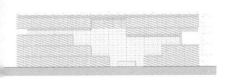

北立面图 | north facade

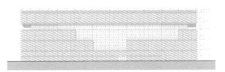

东立面图 | east facade

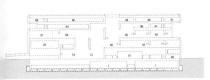

剖面图 | section

剖面图 | section

剖面图 | section

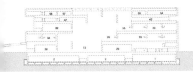

剖面图 | section

1. 基本书库 | basic books stacks
2. 古籍书库 | ancient books stacks
3. 消防水池 | fire water tank
4. 消防水泵房 | fire pump room
5. 中水泵房 | reclaimed water pump room
6. 给水泵房 | water pump room
7. 换热站 | heat exchange station
8. 变电站 | transformer substation
9. 配电间 | power station
10. 值班室 | duty room
11. 空调机房 | air-conditioned room
12. 前厅 | lobby
13. 大厅 | hall
14. 综合咨询台 | comprehensive information desk
15. 公共目录查询区 | public directory query area
16. 办证处 | approval office
17. 电梯厅 | elevator hall
18. 存包处 | luggage office
19. 餐厅 | dining room
20. 厨房 | kitchen
21. 残疾人阅览室 | disabled reading room
22. 书香缘 | Shuxiangyuan store
23. 展厅 | showroom
24. 咖啡厅 | cafe house
25. 报告厅 | lecture hall
26. 视障读者阅览室 | reading room for visually impaired readers
27. 读者自习室 | reader self-study room
28. 自助还书处 | self-service book return
29. 少儿阅览区 | children's reading area
30. 贵宾室 | VIP room
31. 办公室 | office
32. 中文期刊阅览区 | Chinese periodicals reading area
33. 中文图书借阅区 | Chinese book borrowing area
34. 中文报纸阅览区 | Chinese newspaper reading area
35. 读书平台 | reading platform
36. 外部擂台 | external ring
37. 音乐图书馆 | music library
38. 媒体制作室 | media production room
39. 休息室 | resting room
40. 会议厅 | meeting room
41. 多媒体演示室 | multimedia demonstration room
42. 视听文献服务区 | audio-visual document service area
43. 视听室 | audio-visual room
44. 数字资源服务区 | digital resource service area
45. 政府信息查询中心 | government information inquiry center
46. 检索室 | search room
47. 服务中心 | service centre
48. 基本藏书阅览区 | basic library reading area
49. 图书馆学情报学资料室 | library science and information science reference room
50. 网络工程演示室 | network engineering demonstration room
51. 网络工程培训室 | network engineering training room
52. 专题研究室 | special research room
53. 研讨室 | seminar room
54. 繁体中文图书阅览区 | complex Chinese book reading area
55. 外文图书借阅区 | foreign language book borrowing area
56. 地方文献阅览区 | local literature reading area
57. 历史文献阅览区 | historical literature reading area
58. 古籍珍本展室 | rare ancient books exhibition room
59. 天津市古籍保护中心 | Tianjin ancient books protection center
60. 国家古籍修复中心 | national ancient books restoration center

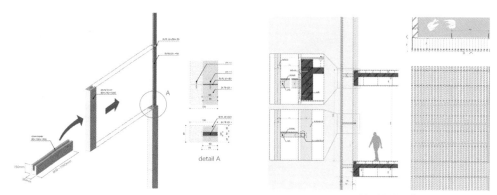

石材百叶构造节点 | stone louver construction

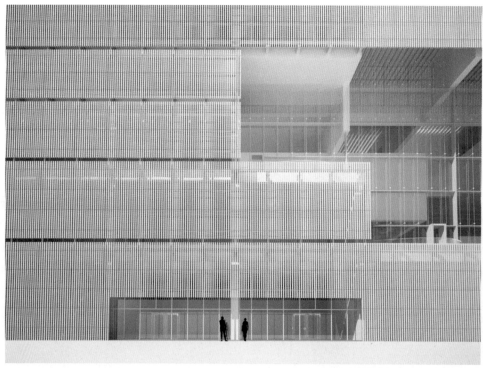

石材百叶构造节点 | stone louver construction node

局部立面 | partial facade

室内实景 | interior scene

室内实景 | interior scene

室内实景 | interior scene

浮在空中的书架 | bookshelf floating in the air

阶梯状阅览平台 | terrace reading platform

阶梯状阅览平台 | terrace reading platform

浮在空中的书架 | bookshelf floating in the air

立体城市空间 | three-dimensional urban space

立体城市空间 | three-dimensional urban space

空中庭院 | air courtyard

空中庭院 | air courtyard

开放的阅览空间 | open reading space

白色顶棚 | white ceiling

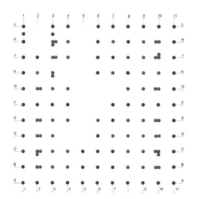
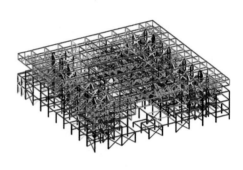

"钢框架支撑旋转交错桁架"体系 | "Steel Frame Supported Rotating Staggered Truss" system

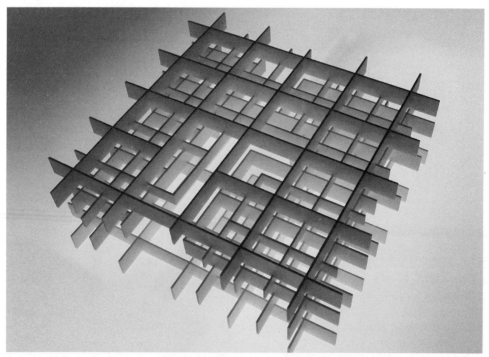

"钢框架支撑旋转交错桁架"体系 | "Steel Frame Supported Rotating Staggered Truss" system

实景 | scene

东南角外观 | southeast corner appearance

随着中央电视台总部大楼的落成,城市公共场所中具有视觉冲击力的建筑形象成为"SUPER 荷兰"的象征,但现在来看那只是沧海一粟。来自荷兰的 MVRDV 建筑设计事务所将居住功能与市场商业功能混合,在共有基地上建立民众喜闻乐见的建筑,使依靠单纯巨构形态的标志性被终结。功能混合、业态杂糅的非传统类型学的作品给荷兰人提供了务实且独特的解决城市疏离问题的方法,这恰恰与滨海文化中心试图搭建以商业为平台、以文化为主角、具有复合功能的市民中心的初衷不谋而合。

滨海图书馆是我们与 MVRDV 建筑设计事务所联合完成的继天津图书馆文化中心馆之后的第二个图书馆项目,它是向文化长廊开放的分享知识的场所。图书馆承担的图书收藏、整理、编辑和借阅的功能已经向信息的收集、整理、交换、发布的本质回归,这是我们设计天津图书馆文化中心分馆时的收获和启示。在天津图书馆创造"智慧乐园"的尝试之后,我们想要在滨海图书馆上向前一步,信息时代需要图书馆由原来单一图书阅读空间向提供一个以知识信息获取为背景的公共共享空间演进。

Binhai Library 滨海图书馆

外立面图 | facade

用地面积	6600 m²
建筑规模	33 700 m²
建筑团队	MVRDV 建筑设计事务所、天津市城市规划设计研究总院、天津中天建都市建筑设计有限公司
结构顾问	三江钢结构设计有限公司、天津大学
幕墙顾问	英海特工程咨询（北京）有限公司
绿建顾问	天津市建筑设计院 – 绿色建筑机电技术研发中心
内装设计	天津市建筑设计院
泛光照明设计	华谊建源照明设计有限公司
智能化设计	天津市建筑设计院 – 绿色建筑机电技术研发中心
获奖情况	2020 年度全国优秀工程勘察设计行业一等奖 2019 年度行业优秀勘察设计奖 优秀公建设计一等奖 2019 年度"海河杯"天津市优秀勘察设计绿色建筑设计二等奖

区位图 | location

"滨海之眼"的概念最初源自眼睛和书本的结合，但由于可行性、整体性问题，经协商将"眼睛"移入室内，形成"眼眶、眼球、书山"的复合意象。这是针对一个概念用一种带有强烈可识别性的形式来表达，再结合地域建造的可行性，在建筑应对城市担负责任的意识指导下，由中外建筑师共同探索其普遍性和秉持共同价值观的直接理性思考的结果。"眼球"经我们的努力成为"会说话的眼睛"，实现发布资讯的功能，使整个共享空间趋近信息传播场所的本质。"书山有路勤为径"是将中国传统的求知方式具象化后通过包络形式操作与"眼球"完美契合，同时使场所更具有体验性、趣味性和舞台效果。

滨海图书馆在 2017 年 10 月开馆以来，迅速成为现象级"网红"图书馆，不只在天津、在中国，甚至在国际上也被广泛关注，荣登美国《时代周刊》发布的"2018 年最值得去的 100 个地方"榜单。这些现象折射出人们普遍地对新复合型城市公共空间的渴望和对单一空洞空间的失望，传达出功能与形式相对应的、理论上的现实局限性和不确定性，昭示着人类追求精神满足的形式超越了物质的枷锁，伴随技术的发展存在无限可能。

With the completion of CCTV Building, the visually striking architectural image of urban public places has become a symbol of "SUPER Netherlands", but now it is only a brief moment of history. MVRDV mixes residential functions with market business functions, and the buildings that people build on the shared base are endowed with the iconic nature of simple giant form. The combination of functions and non-traditional typology of the industry provides the Dutch with a pragmatic and unique approach to solving the problem of urban alienation. This is precisely the case with the Binhai Cultural Center, which is trying to build a civic center with a commercial function as a platform and a cultural function. The original intention coincides.

Binhai Library in Binhai Cultural Center is the second library project after the Tianjin Library in Tianjin Cultural Center, which was jointly completed with the MVRDV. It is a place for sharing knowledge to the Cultural Gallery. The function of the library as a collection, sorting, editing and lending of books has returned to the essence of collecting, sorting, exchanging and publishing information. This is the harvest and inspiration of our design of the Library in Tianjin Cultural Center. After the Library in Tianjin Cultural Center's attempt to create a "smart paradise", we need to take a step forward in Binhai Library. The information age requires the library to evolve from a single book reading space to a public shared space with knowledge information as its background.

The concept of "Eye of Binhai" originated from the combination of eyes and books. However, due to feasibility and holistic problems, the "eyes" were moved into the room through consultation, forming a composite intention of "eyes, eyeballs, and book mountains." This is to express a concept in a form with strong recognizability, combined with the feasibility of regional construction, and under the awareness of the building to respond to the city's responsibility, Chinese and foreign architects explore their universality and share common values. The result of direct rational thinking. The "eyeball" has become the "talking eye" through our efforts to realize the function of publishing the information, so that the entire shared space approaches the essence of the information dissemination place. "The book mountain has a road-to-path" is to visualize the traditional Chinese knowledge-seeking method and then cooperate with the "eyeball" through the envelope form operation, and at the same time make the place more experiential, interesting and stage effect.

Since its opening in October 2017, the Binhai Library has quickly become a phenomenal "internet celebrity" library. It has been widely watched not only in Tianjin, China, but also in the world. It was rated as the one of the mostly wanted visited 100 buildings in 2018 by *Time*. These phenomena reflect the general desire for the new composite urban public space and the disappointment of a single hollow space. It conveys the realistic limitations and uncertainties of the theory corresponding to function and form, indicating that human pursuit of spiritual satisfaction. Forms transcend the shackles of matter, and there is an infinite possibility with the development of technology.

滨海之眼 | eye of binhai

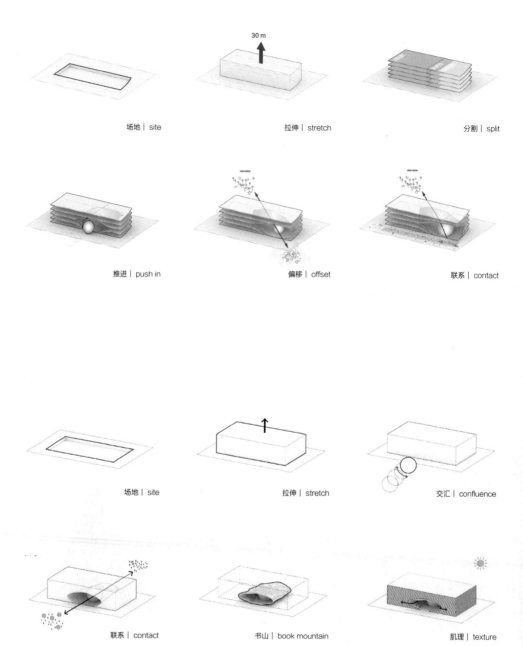

场地 | site   拉伸 | stretch   分割 | split

推进 | push in   偏移 | offset   联系 | contact

场地 | site   拉伸 | stretch   交汇 | confluence

联系 | contact   书山 | book mountain   肌理 | texture

交汇 | confluence

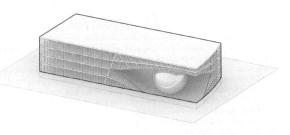

投标方案 | tender scheme

互动 | interaction

推进 | push in

实施方案 | implementation scheme

天光 | sunlight

东立面图 | east facade

左图：图书馆与文化长廊 | library and cultural corridor
右图：东立面图 | east facade

161

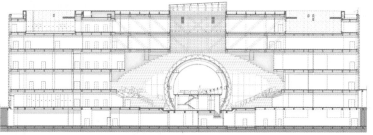

剖面图 | section

施工中的中庭 | atrium under construction

观看区域 | watching

思考区域 | thinking

互动区域 | interaction

书山尺度与行为模式 | scales and behavior of book mountain

中庭 | atrium

球形演示厅表皮最终样块对比 | epidermis comparison

书山与球形演示厅 | book mountain and display hall

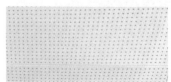

球形演示厅表皮最终样块 | final epidermis

采光井仰视图 | bottom view of light well

球形演示厅表皮外部样块 | epidermis of display hall

正在传递信息的球形演示厅 | display hall

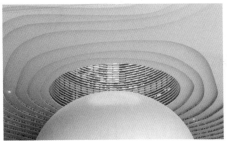

采光井仰视图 | bottom view of light well

采光井正视图 | plan view of light well

检索大厅 | information search hall

阅览室 | reading room

阅览室 | reading room

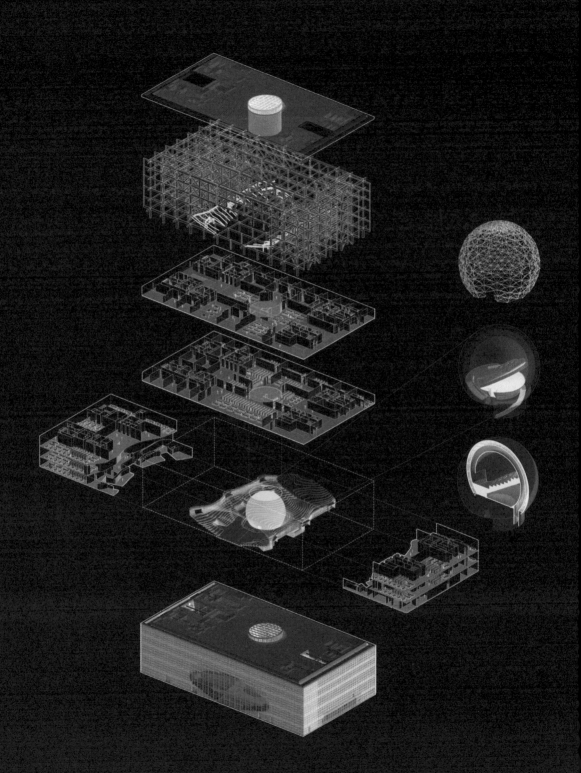

爆炸图 | exploded diagram

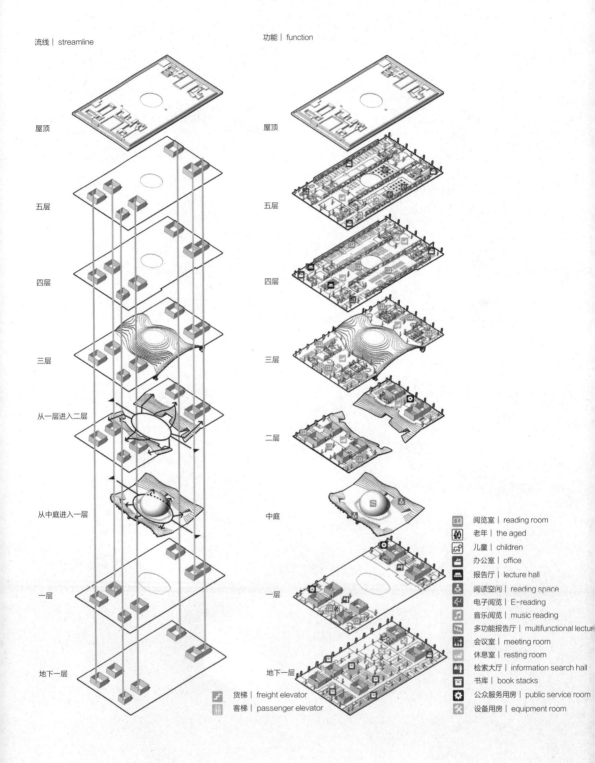

地下一层平面图 | ground floor plan

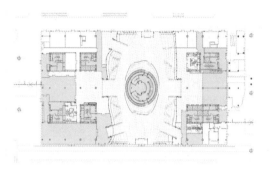

一层平面图 | first floor plan

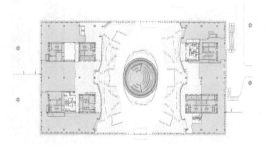

二层平面图 | second floor plan

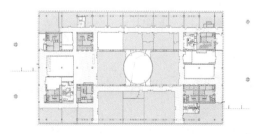

五层平面图 | fifth floor plan

放眼世界的中庭 | atrium

天津塘沽的"红三角"牌纯碱是近代中国民族工业中自主研发并获得国际金奖的产品，它是民族工业的骄傲。作为纯碱生产基地的厂区已不复存在，原址上矗立的两座锥形混凝土冷却塔成为曾经荣耀的唯一记忆，不久也要被拆除了。滨海文化中心最终被布局在此，对于如何传承工业文化遗产以及创造新场所精神给设计师提出了严肃的课题和历史性挑战。

《曼哈顿手稿》是伯纳德·屈米的代表著作，而法国拉·维莱特公园方案的获胜使他的空间形成催生分类活动的概念，挑战了巴黎公共建筑的传统价值，他怀疑惯常性和空间秩序并声称形式与空间中的时间没有固定的联系。这种概念使科技馆方案从开始就呈现出规则形体中插入锥形体块的形象，直率地表达场所记忆与新空间之间的复杂与矛盾。

最初的方案是 7 个大小不一的圆筒自由地植入长方形体块中，虚实体块在内部相互转换并时有交错，随机的位置和偶发的碰撞打破了惯常的秩序，空间的变化隐喻曾经的历史和事件，就如同拉·维莱特公园的立体版，

滨海科技馆
Binhai Science and Technology Museum

建筑面积：32 730 m²
建筑团队｜美国伯纳德·屈米建筑师事务所、天津市城市规划设计研究总院
结构顾问｜三江钢结构设计有限公司、天津大学
幕墙顾问｜英海特工程咨询（北京）有限公司
绿建顾问｜天津市建筑设计院－绿色建筑机电技术研发中心
内装设计｜天津市建筑设计院
泛光照明设计｜华谊建源照明设计有限公司
智能化设计｜天津市建筑设计院－绿色智能建筑设计中心
策展设计团队｜北京众邦展览有限公司
获奖情况｜2019年度行业优秀勘察设计奖优秀公建设计一等奖
　　　　　2019年度"海河杯"天津市优秀勘察设计建筑工程公建一等奖
　　　　　2019年度"海河杯"天津市优秀勘察设计绿色建筑设计二等奖

实景图｜scene

区位图｜location

向外界传递着基地所蕴含的历史能量。经过多位专业人员的讨论及结构可行性分析，最终7个圆筒调整为垂直放置互为图底关系，竖向选择性贯通，并突出屋面形成屋顶特色花园。各层平面按"看见、发现、探索、创造"的路径完成具有现实意义的科技之旅。中央圆筒下圆直径30 m、高40 m，贯通四层直达屋顶，顶部开设直径16 m天窗将光线引入，圆筒中央放置当今世界最大的运载火箭模型，展品与空间相得益彰。

科技馆作为城市文化实践的发生器，在完成对科技的展示及推广的同时，其意义不单纯是历史遗迹表象的保留与再现，更重要的是对功能之外的形式意义的追求与创造，为城市提供一种纪念传统工业文化遗产更为开放共享的表述方法，也许科技馆的建成就预示着新文化遗产的诞生。

"Red Triangle" soda brand in Tanggu, Tianjin, is a product independently researched and developed in the Chinese national industry and won the international gold medal. It is the pride of the national industry. The plant site, which is a soda ash production base, has ceased to exist. The two conical concrete cooling towers on the original site have become the only memories of glory, and are said to have been demolished. The Binhai Cultural Center was finally laid out here, and the planners and designers were faced with serious questions and historical challenges on how to inherit the industrial cultural heritage and create a new place spirit.

The Manhattan Manuscript is the summery of Bernard Tschumi's thinking on conceptual architecture, and the victory of the "La Villette Park" program in Paris has made his space a concept of classification activities that challenged the traditional values of Paris public construction. He doubted the habituality and the spatial order and claimed that the form has no fixed connection with the time in space. This concept makes the science and technology museum program from the beginning to present the image of the cone shape inserted into the regular shape, and express the complexity and contradiction between the place memory and the new space.

The original scheme consisted of seven cones of different sizes that were freely implanted into the rectangular volume. The virtual solid blocks were internally converted and interlaced, and random positions and occasional collisions broke the usual order. The history and events of the past are like the three-dimensional version of "La Villette Park", which conveys the historical energy contained in the base to the outside world. After a multi-disciplinary discussion and structural feasibility analysis, the final six cones were adjusted to be vertically placed in a top-to-bottom relationship, with vertical selective penetration and prominent roofing to form a roofed characteristic garden. Each layer of the plan completes a realistic technological journey by following the path of "seeing-discovering-exploring-creating". The central cone is 30m in diameter and 40m high. It has four floors to the roof. The 16m diameter skylight is used at the top to introduce sky light. The world's largest carrier rocket model is placed in the center of the tube. The exhibits and the space complement each other.

As a science and technology museum of urban cultural practice generators, while completing the display and promotion of science and technology, its significance is not simply the preservation and reproduction of historical relics, but more importantly, the pursuit and creation of formal meaning beyond transcendence. To provide a more open and shared representation of the traditional industrial cultural heritage of the city, perhaps the construction of the science and technology museum itself declares the birth of a new cultural heritage.

长廊正面立面图 | facade of gallery

碱厂 | site of Tianjin Soda Plant

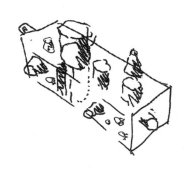

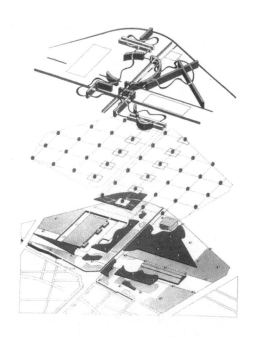

拉·维莱特公园 | La Villette Park

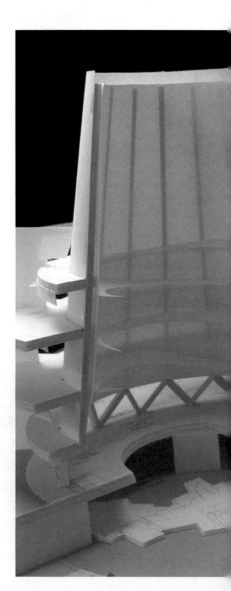

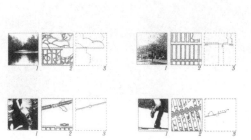

《曼哈顿手稿》内页 | The Manhattan Manuscript

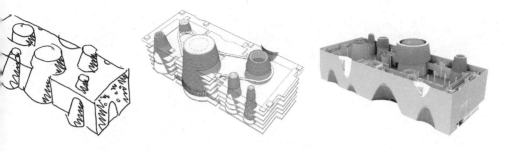

方案呈现 | scheme

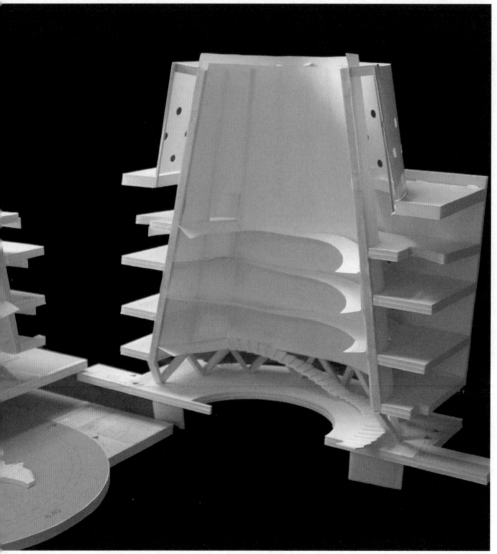

模型照片 | model of photo

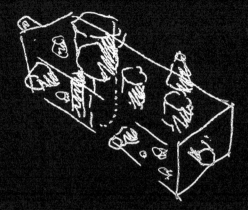
外立面窗孔样式 | perforation style of facade

外立面窗孔样式 | perforation style of facade

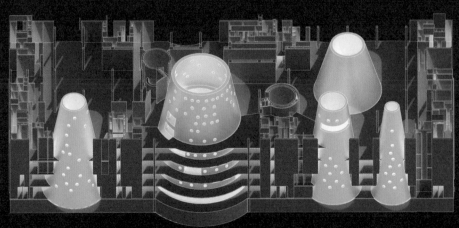
外立面窗孔样式 | perforation style of facade

外立面窗孔样式 | perforation style of facade

效果图 | effect picture

模型图 | model

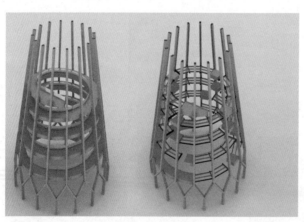

模型图 | model

中央圆筒 | centre cylinder

中央圆筒坡道 | ramp of centre cylinder

筒间流动空间 | flow space in cylinder

筒间流动空间 | flow space in cylinder

筒B一、二层实景图 | cylinder B

玻璃和幕墙分布展开图 | curtain wall expanded

筒间流动空间 | flow space of cylinder

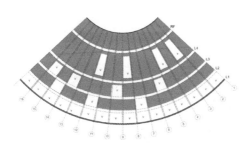

内饰展开图 | expanded view of decoration

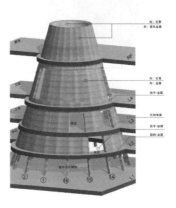

筒 B 一、二层模型图 | cylinder B

筒 B 三、四层 | the third and fourth floor of cylinder B

筒 E | cylinder E

筒 D 三、四层 | the third and fourth floor of cylinder D

外侧实景 | outside scene

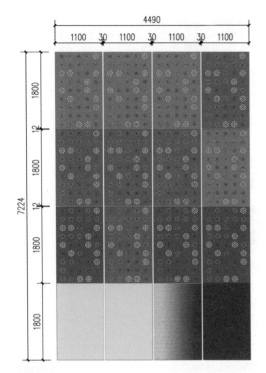
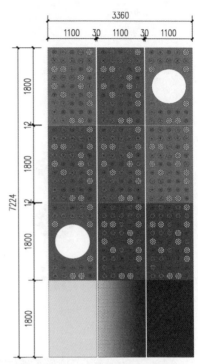

外立面跳色分析 | facade color

外立面图 | facade

外立面穿孔样式 | perforation style of facade

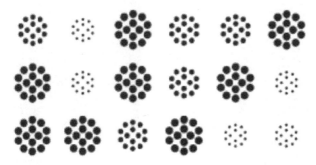

组团穿孔样式 | group perforation style

窗影 | shadow of window

圆窗 | circle window

穿孔样式细部 | detail of perforation style

中央锥形筒结构封顶 | structure of cylinder finished

中央锥形筒内部装饰完工 | decoration of centre cylinder finished

中央锥形筒结构 | structure of centre cylinder

中央锥形筒结构 | structure of centre cylinder

中央锥形筒结构 | structure of centre cylinder

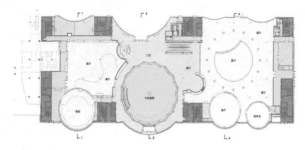

一层平面图 | first floor plan

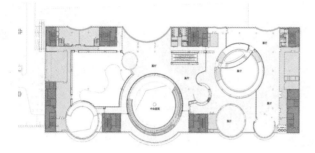

二层平面图 | second floor plan

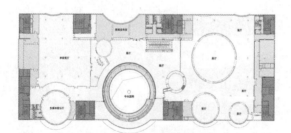

三层平面图 | third floor plan

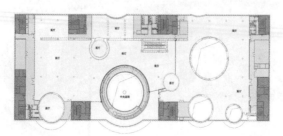

四层平面图 | forth floor plan

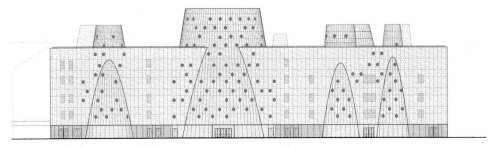

东立面图 | east facade

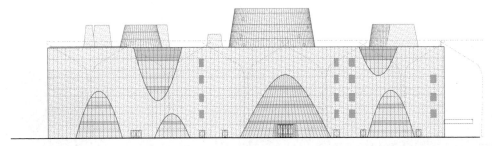

西立面图 | west facade

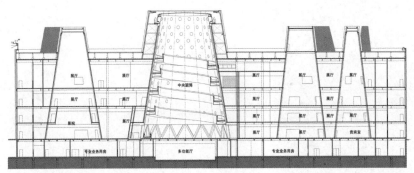

剖面图 2-2 | section2-2

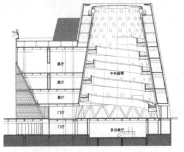

剖面图 3-3 | section3-3

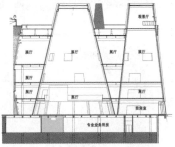

剖面图 4-4 | section4-4

东立面图 | east facade

鸟瞰图 | aerial view

"读四书识五经通六艺"是中国人对于自身文化修养的传统追求,"六艺"是"礼、乐、射、御、书、数"的简称,在古代"六艺"的习得是衡量一个人修养高下的标准。群众艺术馆是现代"六艺"学习、传播的场所,在当下,它的建设为构建现代文明、再塑精神家园提供了良好的机遇。

群众艺术馆的设计理念源自《论语》的"志于道,据于德,依于仁,游于艺",它阐述了中国人的精神追求和生活处事的准则,"游于艺"是培养知识广博和乐活人生的基础。设计中我们用现代建造技术创造空间的开放流动性意象,尝试将"游于艺"的抽象概念用可感知的现实空间体验加以诠释。

我们通过对任务书的分析和讨论,将各种功能按照开放的程度进行分类,并对相关联的功能进行合并。首层布置展示地方戏曲文化的百姓剧场以及"非遗"陈列,二、三层布置艺术培训和部分业务管理用房,四层布

济宁市群众艺术馆
Jining Mass Art Museum

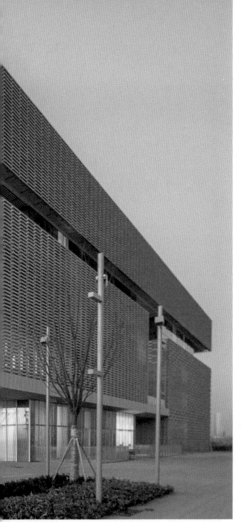

用地面积	济宁文化中心一期可用地面积约 190 310 m²
建筑规模	32 163 m²
建筑团队	天津市城市规划设计研究总院
结构顾问	三江钢结构设计有限公司、美国 H&J 国际公司
幕墙顾问	澳昱冠工程咨询（上海）有限公司
绿建顾问	山东省建筑科学研究院有限公司
内装设计	天津市城市规划设计研究总院
获奖情况	2021 年度"海河杯"天津市优秀勘察设计建筑装饰设计一等奖
	2020 年度"海河杯"天津市优秀勘察设计建筑工程公建一等奖
	2020 年度 BIM/CIM 应用成熟度创新大赛优秀奖
	2019 年度"海河杯"天津市优秀勘察设计 -BIM 应用设计二等奖

实景 | scene

区位图 | location

置职工文体活动、观景平台等，地下一层布置设备、业务及辅助管理用房。除设置两处竖向交通电梯之外，设计中预设一条人们可以进行多种选择的通向各种场所的"活力环线"，环线串接了一个个活力发生器，当人们在环线上移动时，各场所的场景都可以被不同程度地感知，场所之间用"透明"的边界围合，"活力环线"为人们提供了进行无障碍交流的可能。沿"活力环线"的精神回归之旅建立了一条传统美德与现代空间联结的纽带，这是对人们渴望逃离纷杂世界和潜心追求艺术的心理呼应。

馆中，朝西的巨大景框将太白湖的四季之美引入室内，自然美景与不同空间承载的"艺"如绵延画卷缓缓展开，在步移景异的过程中体会"游于艺"的上乐趣和文化熏陶，见证了儒家精神气质的最高境界，在不知不觉中渗透到对空间与环境的体验之中。群众艺术馆是一座根植于传统文化，追求现代空间表达，面向未来的中国本土现代建筑。

"Reading the Four Books and the Five Classics, understanding the Six Arts" is the traditional pursuit of Chinese culture. "Six Arts" is the abbreviation of "Rites, Music, Archery, Riding, Writing, Arithmetic." In the ancient, "Six Arts" acquisition is a measure of the standard of human cultivation. The Mass Art Gallery is the place where modern "Six Arts" spreads. It is the best opportunity for people to rebuild modern civilization and reshape spiritual homes when people are full of whispers.

The design concept of Jining Mass Art Gallery is derived from the sentence "Set your heart on the truth, hold to virtue, lean upon human-heartedness, seek relaxation and enjoyment in the Six Arts" in *Analects of Confucius* which expounds the Chinese people's "spiritual thinking and life affairs" guidelines. The concept of "seek relaxation and enjoyment in the Six Arts" is the foundation for cultivating knowledge and living a happy life. In the design, we use modern construction technology to create the intention of open space, and try to interpret the abstract concept of "walking in art" with the perceptible real space experience.

Through the analysis and discussion of the task book, we classify various functions according to the degree of openness and merge the related functions. The first floor is to arrange the theater of the people who show the local opera culture, the non-legacy exhibition, the second and third floor layout art training and part of the business management room, the fourth floor layout of the staff cultural and sports activities, the viewing platform, etc., the ground floor layout equipment, business and auxiliary Management room. In addition to setting up two vertical traffic elevators, the design presupposes a "dynamic ring" that allows people to make a variety of choices to various places. The loops connect a series of vitality generators when people move on the ring line. The scenes of each place can be perceived to different degrees, and the places are surrounded by "transparent" boundaries. The "vibrant ring line" provides people with the possibility of barrier-free communication. The spiritual return journey of the "Environmental Ring" has established a bond between traditional virtues and modern space. This is a psychological response to those who are eager to escape from the confusing world and pursue art.

Going into the pavilion, the huge frame facing the west introduces the beauty of the scenery of Taibai Lake into the interior. The natural beauty and the "art" carried by different spaces are spread out in layers. In the process of stepping into the landscape, people can have a true feeling of "seek relaxation and enjoyment in the Six Arts" and experience the fun and culture. It witnessed the highest realm of Confucian spiritual temperament and infiltrated into the experience of space and environment without knowing it. The Mass art museum is a modern Chinese building rooted in traditional culture, pursuing modern space expression and facing the future.

实景 | scene

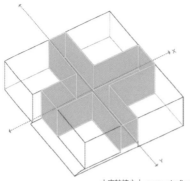

十字轴核心 | cross shaft core

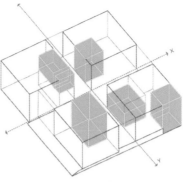

四体位置 | position of four cubes

室外实景 | outdoor scene

室外实景 | outdoor scene

室外实景 | outdoor scene

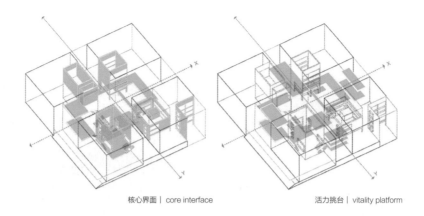

核心界面 | core interface   活力挑台 | vitality platform

中庭 | atrium scene

外幕墙 | curtain wall

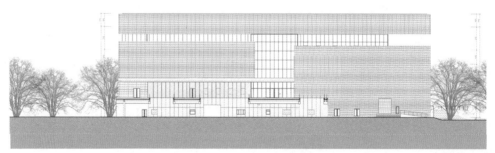

立面图 | facade

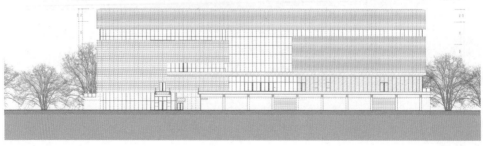

立面图 | facade

外幕墙 | curtain wall

活力环 | vitality loop

活力环 | vitality loop

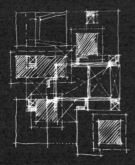

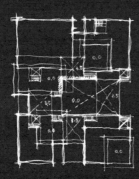

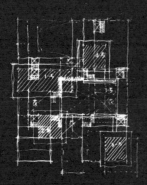

主创手绘图 | main sketch

主创手绘图 | main sketch

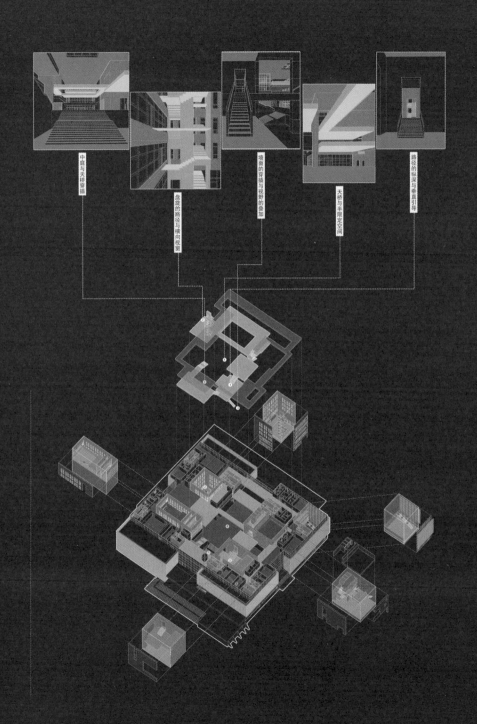

飘板 | floating board

飘板 | floating board

飘板 | floating board

空间渗透 | space penetration

剖面图 | section

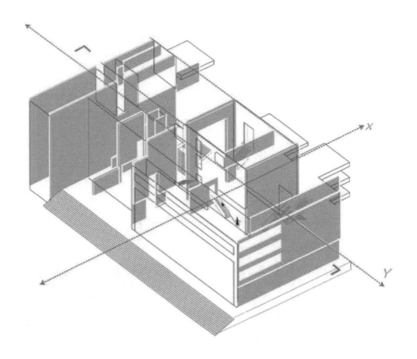

空间圈层化（X 轴） | spatial stratification (X axis)

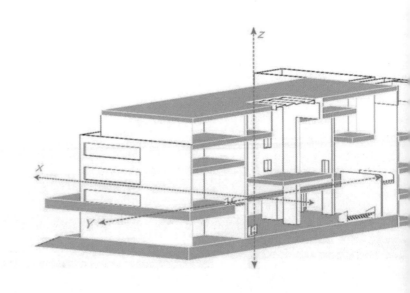

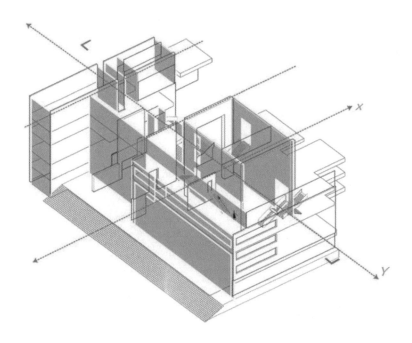

空间圈层化（Y轴） | spatial stratification（Y axis）

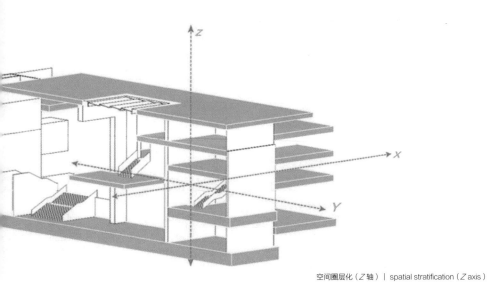

空间圈层化（Z轴） | spatial stratification（Z axis）

一层平面图 | first floor plan

二层平面图 | second floor plan

三层平面图 | third floor plan

四层平面图 | forth floor plan

五层平面图 | fifth floor plan

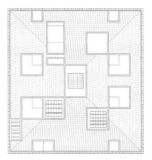
屋顶平面图 | roof plan

从上空看活力环 | vitality loop seen from above

十字轴中庭南向空间 | cross-axis atrium southward space

4m-6m-9m 环线 | 4m-6m-9m loop

导向性清晰的空间活力主环线 | vitality main loop with clear space guidance

9m 挑台回看十字中庭 |
looking at the cross-axis atrium from the 9m high deck

中庭对角空间的引导与强化 |
guidance and enhancement of diagonal space in atrium

中庭 | atrium

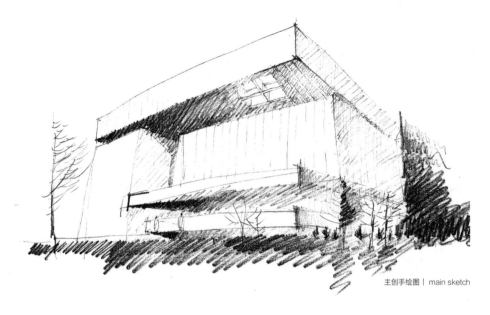

主创手绘图 | main sketch

外立面图 | facade

外檐设计 | exterior wall

群众艺术馆建筑外墙的设计概念来源于济宁非物质文化遗产——鲁锦。我们希望创造一个具有类似传统手工编织质感的幕墙系统，与鲁锦产生形象上的呼应。首先，在肌理构成上，我们通过材质肌理的经纬交织变化打造出建筑的轻盈飘逸与亲切感。其次，通过在建筑同一部位改变石材抛光面与火烧面的颜色及材质平整度，营造粗中有细的形象质感。而建筑具有双层表皮，拙朴的石材与清透的玻璃交相辉映，通过夜景灯光的精妙设计变幻出绚丽的色彩，朦胧中形成一种"织锦"效果。我们希望运用济宁地域性特色材质原料，提炼民俗制作工艺进行演变，在尊重、回应传统的基础上持续创新。

外立面图 | facade

The design concept of the exterior wall of the Mass Art Gallery comes from the intangible cultural heritage of Lu brocade. The architects intend to create the curtain wall system similar to tradition hand-woven texture, which echoes Lu brocade's image. In terms of texture composition, it creates a sense of lightness and intimacy through the interweaving of warp and weft of the material texture. The texture shows as "refined in rough way" by changing the color and material flatness of the stone polishing surface and burning surface in the same part of the building. While the building has a double surface, the rough stone and clear glass send out gorgeous color and create brocade effect under the exquisite design of the night lighting. The architects use the raw materials with regional characteristics in Jining. It is an evolution through refined Folk craft and innovation on the basis of respecting and responding to traditions.

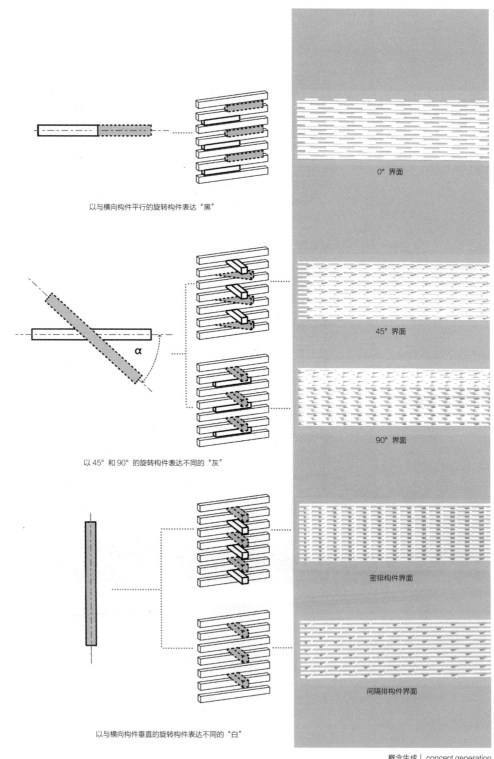

以与横向构件平行的旋转构件表达"黑"

0°界面

以45°和90°的旋转构件表达不同的"灰"

45°界面

90°界面

密排构件界面

间隔排构件界面

以与横向构件垂直的旋转构件表达不同的"白"

概念生成 | concept generation

实与虚 | solid and empty

实与虚 | solid and empty

施工现场 | under construction

施工现场 | under construction

施工现场 | under construction

效果图 | effect picture

实景 | scene

设计团队名单 | Team Members

天津市城市规划设计研究总院建筑设计一院

建筑设计一院现有员工 100 余人，其中国家一级注册建筑师、注册工程师 30 余人。近年来，建筑设计一院依托总院优势，实现了"全过程设计""全流程咨询""全方位把关"的全产业链模式，推进重大公共项目落地。提供从城市设计、前期定位、开发策划、规划、建筑、环境景观等全方位、全流程的工程解决方案。同时，推行"总师责任制"，将先进的设计理念及管理模式贯穿始终，保证设计思想的高质量落地。

建筑一院团队成员（按姓氏拼音排序）

安志红、陈伟杰、陈旭、崔栋、崔磊、邓超、狄阙、董天杰、窦金楠、高艳、宫田芮、郭金未、郭鹏、郭宇、韩海雷、韩佳昕、韩立强、韩宁、侯勇军、胡婷婷、黄波、黄轩、姜雪、解广锋、金彪、康程杰、类成诚、李超、李传刚、李津澜、李钧、李丽媛、李显、李欣艳、李兴宇、李绪良、李雪成、李雅洁、刘畅、刘德忠、刘建红、刘磊、刘亮、刘瑞平、刘月、卢洪权、毛晓亮、孟沫、莫旭昭、牛筱甜、潘宏达、潘林、齐思彤、邱雨斯、曲小美、饶晖、任艳琴、佘江宁、沈伦、石光、苏渊、孙科章、田明、田轶凡、万雪梅、王斌、王立剑、王连顺、王文昌、王洋、王智聪、吴宏婷、吴书驰、吴振兴、邢哲、徐礴骁、闫菁清、闫艺、殷利、袁征、张江铭、张宁、张润兴、张愈芳、赵彬、赵春水、赵堃、郑秀全、周娜

泰安道二号院办公楼
建筑设计： 天津市城市规划设计研究总院
主持建筑师：赵春水
＜设计团队＞
建筑方案：赵春水、张润兴、马凯、张楠楠、田园、
　　　　　李津澜、田垠、邱雨斯
建筑专业：田垠、阳建华、王玫、曲小美
结构专业：崔建敏、何率真、孙受益、饶晖
暖通专业：万雪梅、刘磊
电气专业：贺纯杰、金彪、郭鹏
排水专业：王瑾、张昕

丽思·卡尔顿酒店（泰安道四号院）
建筑设计： 天津市城市规划设计研究总院
主持建筑师：赵春水
＜设计团队＞
建筑方案：赵春水、张润兴、马凯、张楠楠、田园、
　　　　　李津澜
建筑专业：赵春水、王连顺、田垠、阳建华、张胜强、
　　　　　李超、赵彬、田明
结构专业：韩宁、曹学军、于宗志、何率真、饶辉、祁扬、
　　　　　李雪、张明
暖通专业：安志红、万雪梅、高艳
电气专业：贺纯杰
排水专业：王瑾

天津职业大学
建筑设计： 天津市城市规划设计研究总院
主持建筑师：赵春水
＜设计团队＞
建筑专业：赵春水、董天杰、杨怡、李津澜、邱雨斯、
　　　　　田垠、郭宇、陈旭、孟沫、吴书驰、崔磊、
　　　　　刘瑞平
结构专业：饶晖、祁扬、常凤岗、吴振兴
暖通专业：刘磊、安志红、潘林、万雪梅
电气专业：金彪、郭鹏、毛晓亮、吴宏婷
排水专业：王瑾、王斌

天津市民族文化宫
建筑设计： 天津市城市规划设计研究总院
主持建筑师：赵春水
＜设计团队＞
建筑专业：赵春水、田垠、董天杰、吴书驰、史永奇、
　　　　　黄轩、陈伟杰、李欣艳、石光、张泽鑫
结构专业：韩宁、饶晖、祁阳、吴振兴、常凤岗、
　　　　　李雪
暖通专业：潘林、臧效罡
电气专业：金彪、吴宏婷
排水专业：刘德忠

黑牛城道五福里
建筑设计： 天津市城市规划设计研究总院
主持建筑师：赵春水
＜设计团队＞
建筑专业：赵春水、董天杰、杨怡、郭宇、曲小美、
　　　　　刘瑞平、石光、孟沫、田明
结构专业：韩宁、饶晖、李绪良、吴振兴、祁扬、
　　　　　李淑婷
暖通专业：安志红、潘林
电气专业：金彪、吴宏婷
排水专业：王瑾、王斌

天津市第四中学迁址扩建工程
建筑设计： 天津市城市规划设计研究总院、
　　　　　德阁建筑设计咨询（北京）有限公司
主持建筑师：赵春水、雷纳尔·安杰利斯
＜中方设计团队＞
建筑专业：赵春水、侯勇军、陆伟伟、庄子玉、李津澜、
　　　　　田轶凡、赵彬、罗向军
结构专业：张明、孙科章、李绪良
暖通专业：刘磊
电气专业：郭鹏
排水专业：任艳琴
结构咨询：三江钢结构设计有限公司、天津大学

天津图书馆

建筑设计： 株式会社山本理显设计工场、天津市城市
规划设计研究总院

主持建筑师： 山本理显、赵春水

< 中方设计团队 >

建筑专业： 赵春水、王连顺、侯勇军、陆伟伟、张胜强、
崔磊、董天杰、田垠、阳建华、潘福超

结构专业： 韩宁、曹学军、祁阳

暖通专业： 安志红、刘磊

电气专业： 贺纯杰、金彪

排水专业： 杨沛然、张昕

结构顾问： 三江钢结构设计有限公司、天津大学

设备咨询： 天津致通机电设计事务所

滨海图书馆

建筑设计： MVRDV 建筑设计事务所、天津市城市
规划设计研究总院

主持建筑师： 威尼·马斯 (Winy Maas)、赵春水

< 中方设计团队 >

建筑专业： 赵春水、侯勇军、田垠、董天杰、张萌、黄轩、
田轶凡

结构专业： 韩宁、张明

暖通专业： 安志红、刘磊、关键、吴阿蒙

电气专业： 贺纯杰、郭鹏、宋金梁、金彪

排水专业： 武冬宏、任艳琴、杨佩然

照明设计： 张明宇

结构顾问： 三江钢结构设计有限公司、天津大学

设备咨询： 天津致通机电设计事务所

滨海科技馆

建筑设计： 美国伯纳德·屈米建筑师事务所、
天津市城市规划设计研究总院

主持建筑师： 伯纳德·屈米、赵春水

< 中方设计团队 >

建筑专业： 赵春水、董天杰、田垠、侯勇军、田轶凡、
黄轩、韩薇、韩海雷、张萌

结构专业： 韩宁、杨贺先、夏磊、孙科章、张明

暖通专业： 安志红、杨耀东、刘津雅、高艳

电气专业： 何力、刘跃跃、郭鹏

排水专业： 武国增、吕明林、任艳琴

室内设计： 张强、陈平、冯佳、陈乃凡、穆启明、李建波、
张卓

物理环境： 张明宇、刘刚、于娟、田喆

结构咨询： 三江钢结构设计有限公司、天津大学

设备咨询： 天津致通机电设计事务所

济宁市群众艺术馆

建筑设计： 天津市城市规划设计研究总院、
德阁建筑设计咨询（北京）有限公司

主持建筑师： 赵春水

< 设计团队 >

建筑专业： 赵春水、侯勇军、董天杰、李津澜、阳建华、
崔磊、陈旭、刘畅、田轶凡、田园、史永奇

结构专业： 韩宁、杨贺先、饶辉、祁扬、王刚、夏磊

暖通专业： 安志红、刘磊、高艳、杨耀东、刘津雅

电气专业： 金彪、张秋实

排水专业： 王瑾、武国增

内装设计： 赵春水、侯勇军、李津澜、姜辉、崔磊、
刘畅、高艳、林铮迪

结构顾问： 三江钢结构设计有限公司、
美国 H & J 国际公司

幕墙顾问： 澳昱冠工程咨询（上海）有限公司

绿建顾问： 山东省建筑科学研究院有限公司

内装顾问： 厦门俊合建筑设计有限公司

照明顾问： 天津一点照明工程设计有限公司

智能化设计： 天津市城市规划设计研究总院、
天津致通机电设计事务所

照片拍摄者（按姓氏拼音排序）
关永辉、郭鹏、苏振强、田埂、田园、魏刚、邢哲、于果、战长恒、张辉、张明贺、张雨、甄琦

照片提供单位
天津天房酒店管理有限公司丽思卡尔顿分公司、天津市规划和自然资源局、天津市建筑设计研究院有限公司

图纸绘制者（按姓氏拼音排序）
陈伟杰、陈旭、崔磊、狄阙、宫田芮、郭宇、黄轩、康程杰、李传刚、李雅洁、刘畅、刘瑞平、石光、王智聪、吴书驰、薛腾、闫菁清、闫艺、张萌、赵彬、周娜

项目资料提供者（按姓氏拼音排序）
陈伟杰、陈旭、崔磊、郭宇、韩海雷、黄轩、李超、李传刚、李津澜、刘畅、刘瑞平、邱雨斯、石光、田轶凡、田园、吴书驰、张萌、赵彬、周娜